CIRCUIT DESIGN

Techniques for Non-Crystalline Semiconductors

October 9, 2012

Dear Reader,

I sincerely hope this book is useful to you.

I would like to mention that the process of learning is more dynamic than just reading a book. Therefore,

- I will also be posting tutorials and bonus materials on certain aspects I may have browsed through in a non-thorough manner in the book. All this will be hosted on the webpage:

  http://isu.iisc.ernet.in/~ssanjiv/CircuitTechniqueNCS.html

  You can also follow the link to the above page from

  http://isu.iisc.ernet.in/~ssanjiv/Teaching.html

  Therefore, kindly keep tabs on updates on the page. For now, I have posted

  Tutorial 1: How to extract device parameters from a Thin film transistor fabricated in the Lab. The experiments and Discussions

- If you have any queries related to the contents of the book, please feel free to send me a mail at sanjiv.sambandan@gmail.com

- Additionally I would love to receive feedback on what content you feel will be relevant to your work, and I will try my best to help out if I have the expertise in the area.

Since this is the first edition of the book, there is a possibility that errors might have crept in. I will be very grateful if you can get back to me on any typos, confusions in equations, or concepts that you may encounter. I will be more than happy to resolve this immediately.

Thanking You,

S. Sanjiv

CIRCUIT DESIGN

Techniques for Non-Crystalline Semiconductors

Sanjiv Sambandan

CRC Press
Taylor & Francis Group
Boca Raton   London   New York

CRC Press is an imprint of the
Taylor & Francis Group, an **informa** business

CRC Press
Taylor & Francis Group
6000 Broken Sound Parkway NW, Suite 300
Boca Raton, FL 33487-2742

First issued in paperback 2017

© 2013 by Taylor & Francis Group, LLC
CRC Press is an imprint of Taylor & Francis Group, an Informa business

No claim to original U.S. Government works

Version Date: 20120501

ISBN 13: 978-1-4398-4632-2 (hbk)
ISBN 13: 978-1-138-07335-7 (pbk)

Library of Congress Cataloging-in-Publication Data

Sambandan, Sanjiv.
    Circuit design techniques for non-crystalline semiconductors / Sanjiv Sambandan.
      p. cm.
    Summary: "Written for students and professionals within the fields of Materials Science and Engineering, Electronics Engineering, and Applied Physics, this reference provides a systematic means to synthesize circuits with disordered semiconductor field effect transistors (DS-FETs) and explanation of the issues involved. It offers examples on how self-assembly, structural and functional, can be used as a powerful tool in circuit synthesis and provides starting threads for new and future research. The first book to focus on disordered semiconductors, the text covers theory, materials, techniques, and applications, as well as offer practical solutions for semiconductor use in devices"--Provided by publisher.
    Includes bibliographical references and index.
    ISBN 978-1-4398-4632-2 (hardback)
    1. Amorphous semiconductors. 2. Integrated circuits--Design and construction. I. Title.

TK7871.99.A45S26 2012
621.3815'3--dc23
                                       2012014571

Visit the Taylor & Francis Web site at
http://www.taylorandfrancis.com

and the CRC Press Web site at
http://www.crcpress.com

Preface

Disordered semiconductors have been of research interest due to their promise of low temperature fabrication over large areas. The attempt to understand the structure and electronic properties of these materials has generally been driven by the need for a substitute for conventional mono-crystalline silicon for applications in large-area electronics.

There has been significant progress made in the materials and fabrication technologies related to non-crystalline semiconductors. Recent research on complementary polymer semiconductors and fabrication techniques such as ink-jet printing has opened doors to new themes and ideas. However, all these avenues for research meet the hard reality of low mobility and intrinsic time variant behavior in field effect transistors based on these materials. Real world application of these devices in electronic circuits are limited by these fundamental drawbacks. One could argue that the main problem is not mobility or lack of complementary devices, but the threshold voltage shift in the devices that provide a bottleneck in lifetime and does not permit easy analog circuit design.

What is missing is a circuit theory and systematic design approach that would help a designer synthesize circuits without worrying about the nitty-gritty of the physics of the device. In this book, I hope to highlight these problems, provide models, and possible solutions to circuit synthesis with these materials. The book revolves around the problem of threshold voltage shift and the concepts related to the design of electronic circuits around this problem.

This area of research is particularly unique since it brings people from various disciplines — materials science, chemistry, mechanical engineering and electrical engineering — over a common drawing board. This book has been written with the aim of providing the "non-electrical engineer" with the basics and tools that he or she could use to bring their research to the application end and I sincerely hope that this book achieves this goal. While attempting to do so with constraints on time, a shortcoming in the details and exposition to very recent ideas may have appeared. I apologize for this and hope that the book remains useful.

Acknowledgments

I was first formally introduced to non-crystalline semiconductors in my graduate school. Since then, I have learnt a little with every step and this learning has been a cumulative count of the people I have met and the papers and books I have read. Therefore, I would like to thank all the people I have worked with in helping me out with this process of learning.

I would particularly like to thank the following people (arranged in the order I think I met them): Prof. A. Nathan, Dr. K. Sakariya, Dr. A. Kumar, Dr. N. Mohan, Dr. D. Striakhilev, Dr. J. Chang, Prof. A. Sazanov, Prof. P. Servati, Prof. S. Ashtiani, Dr. G. Chaji, Dr. R. A. Street, Dr. R. Apte, Dr. J. P. Lu, Prof. W. Wong, Prof. M. Chabinyc, B. Russo, Prof. A. Arias, Dr. T. Ng, R. Lujan, Dr. A. Pattekar, S. Ready, Dr. G. Whiting, Dr. S. Raychoudhari.

Last but probably the most, I would like to thank my very supporting wife, Prabha, and my parents.

List of Figures

Contents

Part I

Fundamentals

1

Resistor-Capacitor Circuits

CONTENTS

This chapter is dedicated to resistor-capacitor (RC) circuits. In thin film technology, the dynamics of operation of the transistor and circuits composed of transistors can be analyzed using approximate equivalent RC circuits. This chapter provides an introduction to RC circuits and is particularly useful for the reader who is not very familiar with electronic circuits.

First we discuss the resistor and capacitor as circuit elements. Then we discuss concepts related to the dynamics of charging a capacitor, charge sharing between two capacitors, the filtering properties of RC circuits, and finally the energy dissipated in RC circuits.

1.1 The Four Primary Circuit Elements

Passive electrical circuits can be composed of four primary circuit elements - the resistor, capacitor, inductor, and memristor. The four elements are defined by four symmetrical relations between voltage (V), current (I), magnetic flux (Φ), and charge (Q). The four relations are as follows

$$dV = RdI \tag{1.1}$$
$$dV = 1/CdQ$$
$$d\Phi = LdI$$
$$d\Phi = MdQ$$

Here, R is the resistance, C the capacitance, L the inductance, M the memristance. Of these four elements, the memristor was theoretically proposed to exist in 1971 and shown to exist recently [1], [2]. However, as to whether the memristance demonstrated obeys the fourth of the above relations faithfully is debatable.

In this chapter, we concern ourselves with circuits based on resistors and capacitors, as these devices are particularly important in thin film technology.

1.2   Resistor

1.2.1   Resistance and Ohm's Law

The passive resistor is one of the most fundamental of circuit elements. In integrated circuit design resistors are typically fabricated using thin films of metal. Free electrons in a metal behave very much like gas molecules and have a root mean square velocity that scales with temperature such that $v_{rms} = \sqrt{3kT/m_e}$, where m_e is the mass of the electron, k is the Boltzmann coefficient, T is the temperature.

When a potential difference is applied across a metal strip, the free electrons in the metal experience an acceleration due to the electric field. However, the accelerating electron encounters collision with positive ions and other electrons in the metal. The net effect of the continuous accelaration due to a field and subsequent collisions leads to an almost constant velocity of drift of electrons in the metal. The effective drift velocity of free carriers is almost linearly proportional to the applied electric field so that

$$v_d = \mu\xi \tag{1.2}$$

where ξ is the applied electric field, v_d is the drift velocity of electrons, and μ is constant of proportionality defined as the *mobility*. The mobility of free carriers thus defines how "resistant" the metal is to charge transport. The higher the mobility, the lesser the resistance.

The effective movement of electrons through the metal leads to a current. We can now define the *current density* that is the electron current per unit area as

$$J = nqv_d = nq\mu\xi \tag{1.3}$$

where n is the number of electrons per unit volume, and $q = 1.6e - 19C$ is the electronic charge. The total current is therefore given by $I = nq\mu\xi A$, where A is the cross-section area through which the electron flux exists. For small applied voltages V, the electric field can be defined as V/L where L is the length across which the voltage is applied. Therefore, the current through the metal is given by

$$I = \frac{nq\mu A}{L}V \tag{1.4}$$

Thus, macroscopically speaking, the current through a resistor is directly proportional to the applied voltage. This is *Ohm's Law*. The constant of proportionality, is called the *conductance*. The *resistance*, R of the metal strip is the inverse of the conductance and is given by

$$R = \frac{L}{nq\mu A} \tag{1.5}$$

The resistance of a resistor is also called its *impedance* and is measured in *Ohms*, Ω. The current through a resistor and the voltage across a resistor are therefore related as

$$V = IR \tag{1.6}$$

This is the more conventional representation of Ohm's Law.

1.2.2 Resistivity

This definition of resistance is a little uncomfortable since it is dependent on the geometry of the metal strip. Instead we can define a geometry independent property of the material called the *resistivity*, ρ, which is resistance offered to charge transport by a material of unit length and unit cross-section area.

$$\rho = 1/qn\mu \tag{1.7}$$

The resistivity is independent of the dimensions of the resistor and is purely a material property. Since it is dependent on the mobility of the carriers, it is also a function of the temperature. Given a certain geometry of a metallic film, we can now determine and quantify its resistance from the knowledge of the resistivity of the metal used. The resistance and resistivity are related as

$$R = \rho L/A \tag{1.8}$$

If we connect two resistors in series, it is equivalent to increasing L. If we connect two resistors in parallel, it is equivalent to increasing A. Thus, the effective resistance of two resistors of resistances R_1 and R_2 in series is $R_1 + R_2$. On the other hand, the effective resistance of having R_1 and R_2 in parallel is $(R_1^{-1} + R_2^{-1})^{-1}$.

1.2.3   Power Consumption in a Resistor

When current flows through a resistor, the resistor dissipates heat and thereby consumes power. The power consumption, P, in a resistor of resistance R, with a constant current I through it is given by

$$P = I^2 R \tag{1.9}$$

On the other hand, if the voltage applied across the resistor, V is constant, the current through the resistor is V/R. Hence, the power consumption in the resistor is

$$P = V^2/R \tag{1.10}$$

The total energy lost in dissipation through the resistor is $\int_{t=0}^{t=T} P\,dt$, where the power P is consumed for time T. In general, when a voltage $v(t)$ is applied across a circuit element, and if a current $i(t)$ is drawn by the circuit element, the total energy consumed from the power supply from time $t = t_1$ to time $t = t_2$ is $\int_{t=t_1}^{t=t_2} v(t)i(t)\,dt$.

1.3   Capacitor

1.3.1   Capacitance and Charge on a Capacitor

The passive capacitor is an energy storage device while casually speaking it can be termed as a device that stores charge. A capacitor in general consists of a dielectric medium sandwiched between two conductive electrodes. A dielectric is a material with very high resistivity that can be polarized by an applied electric field. Thus a dielectric placed in an electric field does not allow charges to easily flow through it. However, in the presence of an electric field, the electrons in the atoms of the dielectric and the positively charged nucleus shift from their equilibrium position forming a dipole. In a capacitor, the electron cloud of the atoms in the dielectric shift towards the positive electrode, and the positively charged nucleus shift of the atoms shift towards the negative electrode. This is called as *dielectric polarization*. Dielectric polarization creates an internal electric field which opposes the applied electric field thereby reducing the overal field within the dielectric. If the polarizability of the dielectric is very high, the dielectric is said to have a high *dielectric constant*. Thus, by applying a voltage across the two electrodes of a capacitor, one can store charge (and hence energy). The dielectric constant of a dielectric refers to its relative permittivity. The relative permittivity of a material is the ratio of its absolute permittivity, ϵ, to the permittivity of vacuum, and is represented as ϵ_r or k (not to be confused with Boltzmann's constant). It must be noted that the permittivity is dependent on the frequency of measurement (and operation) and in this book we always imply the static or low frequency permittivity by the usage of the term. The permittivity of vacuum is $\epsilon_0 = 8.85e - 12F/m$. The dielectric constants of vacuum, air, silicon dioxide, silicon nitride, and most non polar polymer dielectrics are $1, \approx 1, \approx 3.9, \approx 6$, and $\approx 2 - 3$, respectively.

The *capacitance*, C, of a capacitor is the ratio of the change in charge stored to the change in the voltage across the electrodes,

$$C = dQ/dV \tag{1.11}$$

The capacitance, voltage across the passive capacitor, and the charge stored on the capacitor are related as

$$Q = CV \tag{1.12}$$

In order to calculate the capacitance of a capacitor for any random geometry, we apply *Gauss' law* to establish the relation between the charge stored in the capacitor and the voltage across the capacitor due to the charge stored. Gauss' law states that the total electric flux out of a closed surface is equal to the charge enclosed by the divided by the permittivity, i.e.

$$\int_S \xi.dA = \frac{Q}{\epsilon} \tag{1.13}$$

Here, $\int_S \xi.dA$ is the surface integral of the electric field, which is the electric flux, and Q is the total charge enclosed by the surface. Using this law we establish the electric field and from there the potential drop across a capacitor. The relation between the potential drop and the stored charge can be used to evaluate the capacitance using Equation 1.11.

Poisson's Equation

Gauss' law can also be written in differential form. According to the Divergence theorem $\int_S \xi.dA = \int_V (\nabla.\xi)dV$, where $\nabla.\xi$ is the divergence of the electric field, and $\int_V (\nabla.\xi)dV$ represents its volume integral. Therefore, Gauss' Law becomes

$$\nabla.\xi = \frac{\rho}{\epsilon} \tag{1.14}$$

where ρ is the charge density per unit volume. For the one dimensional case (say x direction), Poisson's equation becomes,

$$\frac{\partial \xi}{\partial x} = \frac{\rho}{\epsilon} \tag{1.15}$$

We use this form of Gauss' equation for the analysis of semiconductor devices.

A common capacitor architecture is the parallel plate capacitor, which consists of two rectangular plate like electrodes sandwiching a dielectric between them. If the plates have an overlap area of A, and the distance between the plates is d, the capacitance of the parallel plate capacitor is $C = \epsilon A/d$ where $\epsilon = k\epsilon_0$ with ϵ_0 being the permittivity of vaccuum, and k being the dielectric coefficient. If two capacitors are placed in parallel, then we effectively increase A while if two capacitors are placed in series, we effectively increase d. Therefore, for two capacitors of capacitances, C_1 and C_2 in parallel, the effective capacitance is $C_1 + C_2$. If C_1 and C_2 are in series, the effective capacitance is $(C_1^{-1} + C_2^{-1})^{-1}$.

1.3.2 Current through a Capacitor

Let us now consider a capacitor with one electrode connected to ground. When a certain quantity of charge is placed on the free plate, this immediately results in the other plate taking charge from the ground terminal to compensate for the applied charge. After this there is no more movement of charge. Now, when the charge on one electrode of the capacitor continuously varies with time, it causes charge variation on the other electrode as well. The variation of charge with time, by definition, indicates a current through the capacitor.

This current through the capacitor is given by the rate of change of charge stored in the capacitor, i.e.

$$
\begin{aligned}
I &= dQ/dt \\
&= d(CV)/dt \\
&= CdV/dt + VdC/dt
\end{aligned}
\qquad (1.16)
$$

If the capacitance of the capacitor is constant, the exists a current through the capacitor only when there is a time-varying voltage across its electrodes, i.e., $I = CdV/dt$.

1.3.3 Equivalent Impedance of a Capacitor

The question that we now ask is: what is the impedance of the capacitor? Clearly due to the above definition of current, the current through the capacitor is proportional to the rate of change of voltage. Hence, the impedance or effective resistance of a capacitor is dependent on the rate or frequency of the voltage variation across the plates. If we consider a sinusoidal voltage waveform $v(t) = v_0 sin(\omega t)$ across the capacitor plates, the current is given by $i(t) = Cv_0\omega cos(\omega t)$. The amplitude of the current is therefore $Cv_0\omega$ and the current waveform is phase shifted by 90 degrees. Thus, the magnitude of the effective resistance of the capacitor is given by $1/\omega C$, where $\omega = 2\pi f$ is the cyclical frequency in radians/sec. In order to include the phase shift component, the impedance of a capacitor is a complex quantity $1/j\omega C$.

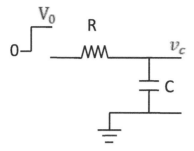

FIGURE 1.1
Series resistor-capacitor circuit.

1.3.4 Energy Stored in a Capacitor

The capacitor is an energy storage device and the energy stored in a capacitor of capacitance C with a voltage V across the plates is given by

$$E = \frac{CV^2}{2} \tag{1.17}$$

1.4 Series Resistor-Capacitor (RC) Circuit

Let us consider a resistor in series with a capacitor, with the free electrode of the capacitor connected to ground shown in Figure 1.1. This circuit, called a *series RC circuit*, is of great importance. Field effect transistor based circuits can usually be broken down into simple RC circuits that can be locally analysed to obtain information useful for design.

1.4.1 Capacitor Charging

Assuming the capacitor is initially devoid of charge, we study the dynamics of charging of the capacitor when a step voltage input of magnitude V_0, is applied to the input of the resistor.

Since the current through the resistor and capacitor is the same, we equate the current through both elements. First, we define a time varying variable, $v_c(t)$, which represents the voltage on the capacitor plate. Then, the current through the resistor can be found to be $(V_0 - v_c(t))/R$ while the current through the capacitor is found to be $C dv_c(t)/dt$. For the sake of brevity, we henceforth represent $v_c(t)$ as v_c. Equating the currents through the resistor and capacitor we obtain

$$V_0 - v_c = RC\frac{dv_c}{dt} \tag{1.18}$$

The solution to this differential equation with the initial condition that $v_c = 0$ at time zero yields,

$$v_c = V_0(1 - e^{\frac{-t}{RC}}) \tag{1.19}$$

The product of the resistance, R and capacitance C defines the *time constant* of the circuit. The time constant of an RC circuit determines how quickly a circuit can be charged. The lower the time constant, the lower the time required to charge the capacitor. Since the capacitor ideally takes infinite time to charge fully, we consider the capacitor charged if the voltage on the capacitor is about 90% of the applied voltage, V_0. This implies that the time taken to charge a capacitor is $\approx 2.3RC$.

In general if a capacitor has some initial voltage $V_i < V_0$, and is charged by a step input voltage V_0, the dynamics of charging is given by

$$v_c = V_0 - (V_0 - V_i)e^{\frac{-t}{RC}} \tag{1.20}$$

1.4.2 Capacitor Discharging

Equivalently we can also consider the problem of discharging a capacitor that has some initial charge V_0. If this capacitor is discharged to some potential V_f via an RC circuit, the dynamics of discharging the capacitor is given by

$$v_c = V_f + (V_0 - V_f)e^{\frac{-t}{RC}} \tag{1.21}$$

1.5 Charge Sharing between Capacitors

Consider a capacitor of capacitance C_1 with one electrode connected to ground, holding a charge Q_{1i}. Clearly, the voltage across this capacitor is $\frac{Q_{1i}}{C_1}$. Now consider another capacitor of capacitance C_2, with no charge and with one electrode connected to the same ground. Figure 1.2 describes this setup. We now connect the two top plates of the two capacitors with a metallic conducting wire. What is the final charge on both capacitors at steady state?

Since the top plates of both capacitors are shorted by the wire, their potential must be the same at equilibrium. Let this final potential be V_f. Now, since the total charge must be conserved, the charge on the individual capacitors must add up to the total charge. Therefore,

$$C_1 V_f + C_2 V_f = Q_{1i} \tag{1.22}$$

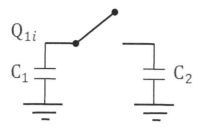

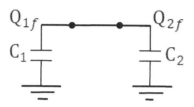

FIGURE 1.2
Charge sharing between two capacitors.

Thus, $V_f = \frac{Q_{1i}}{(C_1+C_2)}$, and the final charge on C_1 and C_2 at steady state is,

$$Q_{1f} = C_1 V_f = \frac{C_1 Q_{1i}}{C_1 + C_2} \qquad (1.23)$$

$$Q_{2f} = C_2 V_f = \frac{C_2 Q_{1i}}{C_1 + C_2}$$

Studying charge sharing using energy balance

There is another way to go about calculating the charge on the capacitor plates by considering the conservation of total energy in the system. In the above example, the total initial energy of the system is the energy stored on C_1, which is $\frac{Q_{1i}^2}{2C_1}$. After C_1 and C_2 are shorted, the energy stored in the two capacitors is $\frac{(C_1+C_2)V_f^2}{2}$. During the transfer of charge via a current through the resistance (say R) of the metallic shorting wire, there was some heat dissipated in the wire. This heat loss is defined as $\int_0^\infty i(t)^2 R\, dt$ where $i(t)$ is the current through the resistor at time t with time zero being the moment we short the two capacitors. If we define $V_0 = Q_{1i}/C_1$, the correct and complete energy balance equation is

$$\frac{C_1 V_0^2}{2} = \frac{(C_1 + C_2)V_f^2}{2} + \int_0^\infty i(t)^2 R\, dt \qquad (1.24)$$

In order to solve this equation, we need to find $i(t)$.

Let the instantaneous voltages on C_1 and C_2 at time t be $v_1(t)$ and $v_2(t)$, respectively. We now have the following set of equations

$$-\frac{v_1(t) - v_2(t)}{RC_1} = \frac{dv_1}{dt} \tag{1.25}$$

$$\frac{v_1(t) - v_2(t)}{RC_2} = \frac{dv_2}{dt}$$

Defining, $\Delta v(t) = v_1(t) - v_2(t)$, and subtracting the second equation from the first, we find

$$-\frac{\Delta v(t)}{R}\left(\frac{1}{C_1} + \frac{1}{C_2}\right) = \frac{d\Delta v(t)}{dt} \tag{1.26}$$

Since $\Delta v(t) = V_0$ at $t = 0$, we solve this equation to find

$$i(t) = \frac{\Delta v(t)}{R} = \frac{V_0}{R}e^{\left(-\frac{t}{R}\left(\frac{1}{C_1} + \frac{1}{C_2}\right)\right)} \tag{1.27}$$

The power dissipated in the resistor is therefore given by

$$\int_0^\infty i(t)^2 R \, dt = \frac{V_0^2}{2\left(\frac{1}{C_1} + \frac{1}{C_2}\right)} \tag{1.28}$$

Substituting for the power dissipation term in the main energy balance equation, Eq. 1.24, we obtain $V_f = \frac{C_1}{C_1+C_2}V_0$ and the final charges on C_1 and C_2 can be found to be the same as that predicted by charge balance in Eq. 1.23.

1.6   Filtering Property of RC Circuits

When a step voltage is applied at the input of an RC circuit, the RC circuit delays the arrival of the step input on the capacitor plate. This delay limits the rate at which the voltage on the capacitor plate can follow an input voltage. Consider a pulse stream input to an RC circuit with the capacitor having zero initial charge as shown in Fig. 1.3. As the pulse goes from 0 to V_0, the capacitor begins to charge to the potential V_0, and when the pulse goes from V_0 to 0, the capacitor begins to discharge through the resistor. Let us assume the pulse has equal on and off time periods. The width of the pulse (on and off cycle) is T. Let us define the capacitor to be fully charged if the voltage

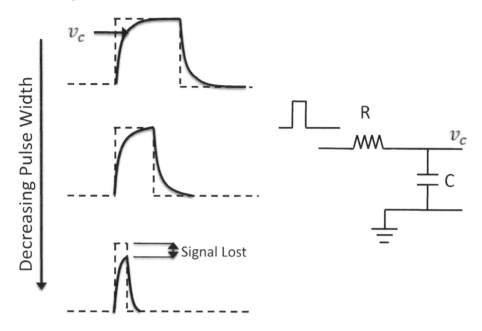

FIGURE 1.3
Filtering Properties of RC circuits.

on the capacitor at the end of the high going cycle, $v_c(T)$ is more than 90% of V_0.

When the pulse width, T is much wider than the charging time of the capacitor $(>> 2.3RC)$, the waveforms of capacitor charging effectively follow the input voltage and charges completely albeit with a delay much smaller than the pulse width. Now let us imagine the pulse width is slowly reduced. As we continuously reduce the pulse width, the capacitor charging voltage is not affected till as long as $T \geq 2.3RC$. However, the moment $T < 2.3RC$, the capacitor does not charge fully. Further reduction in the pulse width also reduces the total charging voltage on the capacitor.

Let us now plot this event. We define the gain of the circuit as $A = \frac{v_c(T)}{V_0}$. On the y-axis we plot $\log_{10}(A)$ and on the x-axis we plot the logarithm of the pulse frequency, which is $\log_{10}(1/T)$. Since we defined the capacitor to be fully charged if it had charged to more than 90% of the applied voltage, we will see that till $1/T < 1/2.3RC$, A remains constant. Beyond this point, the gain A begins to drop.

In this simple experiment one can see the filtering nature of the RC circuit where the lower frequency components were "passed" through with $A \geq 0.9$ and the higher frequency components were "filtered" off due to the reduction in gain. Thus, the RC circuit is basically a *low pass filter*. If a random signal with many frequency components were to be presented at the input of the

RC circuit, the lower frequency components of the signal get passed onto the capacitor with higher gain, while the higher frequency components get passed onto the capacitor with lower gain. Moreover, there seems to be a *cutoff frequency* related to the time constant RC beyond which the gain reduces and before which the gain remains almost constant same.

Strictly speaking, one would carry out an experiment like this with a sinusoidal input of various frequencies and measure the amplitude of the sinusoidal output. The power of the gain is defined as A^2. The plot of $\log_{10}(A^2)$, which is the "output power" in decibels (dB), versus $\log_{10}(F)$ is called a *Bode Plot*. The cutoff frequency is then defined as the frequency at which the output power falls to one half the low frequency value. In the case of RC circuits, this cutoff frequency is $1/2\pi RC$.

1.7   Impedance of the RC Circuit

Before we close this chapter, we define the impedance of the series RC circuit. Since the two elements are in series, the impedance of the RC circuit is the sum of their individual impedances. Thus,

$$|Z| = |R + 1/j\omega C| = \sqrt{R^2 + (1/\omega C)^2} \tag{1.29}$$

The current through the RC circuit is the applied voltage by the effective impedance,

$$|I| = \left| \frac{V_0}{R + 1/j\omega C} \right| = \frac{V_0}{\sqrt{R^2 + (1/\omega C)^2}} \tag{1.30}$$

The voltage on the capacitor, is $|I/j\omega C|$

$$|v_c| = \left| \frac{V_0}{1 + jR\omega C} \right| = \frac{V_0}{\sqrt{1 + (\omega C R)^2}} \tag{1.31}$$

The filtering property of the RC circuit becomes apparent in the above equation. Since $A^2 = \frac{1}{1+(\omega C R)^2}$, it takes a value of $1/2$ when the frequency $\omega = 1/RC$ or $f = \omega/2\pi = 1/2\pi RC$. This frequency is the cutoff frequency of the RC filter.

1.8   Conclusion

This chapter presented the series RC circuit, its properties, and its analysis. This circuit is of fundamental importance since in the future chapters we will return to these concepts in order to analyze transistor-capacitor circuits that form the basis of present-day large-area electronic systems. We will learn how to compute the resistance of the transistor and hence convert all transistor-capacitor circuits to equivalent RC circuits in order to facilitate back-of-the-envelope calculations for quick design. These concepts can be futher studied and analyzed from any textbook in basic electrical engineering.

2

Fundamentals of Semiconductor Devices

CONTENTS

In this chapter we introduce concepts related to semiconductors and semiconductor devices. The primary aim of this chapter is to introduce the unfamiliar reader to semiconductor device analysis.

2.1 Energy Levels and Energy Bands

The electrons of a single isolated atom of any element occupy atomic orbitals that are assigned certain discrete energy levels. The energy levels are quantized and there are certain energies that the electrons can never possess called the *forbidden energy levels*. If several atoms are brought together into a molecule, their atomic orbitals split. As more and more atoms are brought together, a large number of orbitals are created with the separation in their energy levels becoming smaller. Solids, which are composed of a large number of atoms (about 10^{23}), thus have a large number of orbitals occupying a large number of finely separated energy levels called *energy bands*. The forbidden energy levels form *energy band gaps* between bands.

At 0K temperature, all the energy levels associated with the electrons bound to a nucleus are fully occupied. The highest band of these occupied energy levels constitutes what is called the *valence band*. When the electrons absorb energy (at higher temperatures, due to interaction with high energy photons, etc.) and escape the nucleus of the atom, they are free to move in material. The band of energies the free electrons may occupy is defined as the *conduction band*. Electrons in the conduction band are responsible for electric conduction in solids. The conduction band is separated from the valence band by an energy gap.

2.2 Metals, Semiconductors, Insulators

What makes a metal different from an insulator? How do we define semiconductors? The answer to these questions lie in the size of the band-gap. Under the condition of thermal equilibrium, the electrons in the valence band obtain thermal energy from the environment. In the case of insulators, the band-gap is large (about 5 to 6eV or more). Thus, in order to excite electrons from the valence band to the conduction band, we need large amounts of energy which is not available at a standard room temperature of 300K. In the case of semiconductors the band-gap is much smaller (about 2eV). At 300K temperature, the thermal energy transferred to the electrons is sufficient to excite some of them into the conduction band. In the case of metals, the valence and conduction band overlap to form a single band that is partly empty and partly

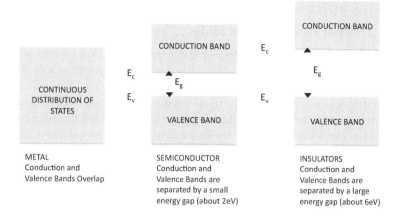

FIGURE 2.1
Band diagram of metals, semiconductors and insulators.

filled. These different classes of materials based on electrical conductivity are illustrated in Figure 2.1. This representation of materials with the conduction and valence bands along with the band gap is called the *band diagram*.

The conductivity of these materials is therefore dependent on the number of carriers in the conduction band, which is in turn a function of the band-gap of the material. The band-gap is thus one of the most influential aspects of the band structure, as it strongly determines the electrical and optical properties of the material. The band-gap and defect states created in the band-gap (by techniques like doping) can be used to create semiconductor devices such as solar cells, diodes, transistors etc.

2.3 Semiconductor Fundamentals

As discussed above, the semiconductor can be modeled with a conduction band and a valence band separated by an energy gap. Both the conduction and valence bands have states, while the energy gap is void of states unless there are defects present in the semiconductor lattice. Thus, charge carriers occupy states in the conduction band and valence band. The top of the valence band is denoted to have an energy E_v while the bottom of the conduction band is denoted to have an energy E_c. The energy gap, E_g, in the semiconductor is therefore defined as $E_g = E_c - E_v$. Using this model of the semiconductor, we study the different aspects of charge transport in these materials.

2.3.1 Carriers

2.3.1.1 Free Electrons and Holes

The electrons present in the valence band correspond to the electrons involved in the bonding and are therefore not involved in charge transport. However, when a bond is broken, the associated electron is free and can contribute to charge transport. These *free electrons* occupy states in the conduction band. The band-gap of the material, directly corresponds to the the energy required to break a bond and thereby releasing a carrier from the valence band to the conduction band. When the bonds of the semiconductor lattice are broken, electrons are pushed to the conduction band, and leave behind vacancies called *holes*.

Now consider a block of semiconductor, with voltage applied across its ends. The free electrons in the conduction band of the semiconductor rush towards the positively charged electrode. The electrons in the valence band, jump in and out of the vacancies as they attempt to be bound by the atoms that are closest to the positively charged electrode which appears as though the void moves in the opposite direction. This is equivalent to the holes in the valence band moving towards the negatively charged electrode. Therefore, we may think of holes as free positively charged particles residing in the valence band, while the free electrons are negatively charged particles residing in the conduction band.

2.3.1.2 Effective Mass

When an electron is placed in vacuum in a uniform electric field, ξ, the velocity of the electron, v, is related to the field as

$$-q\xi = m_e dv/dt \qquad (2.1)$$

where m_e is the mass of the electron. However, when the electrons move in the semiconductor lattice under the presence of an electric field, they interact with the fields generated by the atoms in the semiconductor. In order to fully understand the motion of the electron in these periodic fields, we need a quantum mechanical treatment of the problem. However, the problem can be simplified greatly by considering a macroscopic picture and defining the equation of motion of the electrons in the semiconductor as

$$-q\xi = m_e^* dv/dt \qquad (2.2)$$

where m_e^* is the effective mass of the electron in the semiconductor. Similarly, holes too are identified with an effective mass m_h^*.

2.3.2 Density of States

The *density of states* in the semiconductor defines the number of *states* per unit volume an electron can occupy (which implies the energies that an elec-

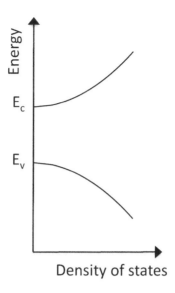

FIGURE 2.2
Nature of the density of states in a semiconductor assuming equal effective masses for electrons and holes.

tron can have), associated with each energy level. The density of states in the conduction band, $g_c(E)$, tells us that there are $g_c(E)dE$ conduction band states found between the energy levels E and $E + dE$, with $E \geq E_c$. On the other hand, the density of states in the valence band, $g_v(E)$ tells us that there are $g_v(E)dE$ states present in the valence band, given $E \leq E_v$. The energy gap which is forbidden for the electrons to occupy does not have any state ideally. From quantum mechanical treatments, the density of states in the conduction and valence bands are given by

$$g_c(E) = \frac{8\pi(2(m_e^*)^3(E - E_c))^{1/2}}{h^3} \quad \text{for} \ \ E \geq E_c \tag{2.3}$$

$$g_v(E) = \frac{8\pi(2(m_h^*)^3(E_v - E))^{1/2}}{h^3} \quad \text{for} \ \ E \leq E_v$$

$$\tag{2.4}$$

Figure 2.2 illustrates the qualitative nature of the shape of the density of states.

2.3.3 The Fermi Level and Fermi Function

The *Fermi function*, $f(E)$, defines the probability of finding an electron at a given energy level E. Equivalently, the probability of finding holes at the energy level E is given by $1 - f(E)$. The Fermi function is therefore a probability density function,

$$f(E) = \frac{1}{1 + e^{\frac{E - E_f}{kT}}} \tag{2.5}$$

where, E_f is the *Fermi level*, k is the Boltzmann's constant, and T is the temperature. By definition, the Fermi level is the energy level at which the probability of finding an electron is exactly $1/2$ at any temeprature. At 0K, the probability of finding an electron below the Fermi level is 1 while the probability of finding an electron above the Fermi level is 0. The Fermi level at 0K can be considered to be representative of the highest level in energy the electron can occupy *if it had a state to occupy at that energy level.*

If $E - E_f >> kT$, the Fermi function can be replaced by the Maxwell-Boltzmann distribution function

$$f(E) \approx e^{-\frac{E - E_f}{kT}} \tag{2.6}$$

If $E - E_f << kT$,

$$1 - f(E) \approx e^{-\frac{E_f - E}{kT}} \tag{2.7}$$

These approximations are useful for quick calculations. However, if $E - E_f$ becomes comparable to or is less compared to kT, these approximations fail and the full Fermi function must be used.

Water Bucket Analogy

The physical meaning of the Fermi level can be described by the following analogy.

Consider an empty bucket of a given shape. The space in the bucket is analogous to the states and the shape of the bucket is analogous to the density of states while the height corresponds to the energy. The energy levels increase in energy as we move from the floor of the bucket to the mouth. As we pour water into the bucket, each water droplet (or molecule or the smallest packet of water) — analogous to the electron — begins to occupy one state after another from the lowers energy level and slowly climbs up. When the water is completely poured into the bucket, the surface of the water in the bucket corresponds to the states in the highest energy level occupied by the electrons. This surface refers to the Fermi level.

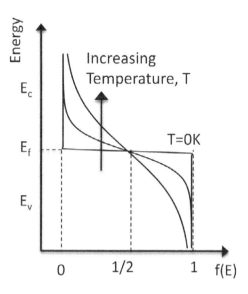

FIGURE 2.3
The Fermi function at different temperatures. The value of the Fermi function is always $1/2$ at the Fermi level.

2.3.4 Number of Carriers

From the definition of the density of states and the Fermi function, we can estimate the number of carriers occupying a certain energy level at any given temperature. The number of electrons at any given energy level is the product of the number of states available at that energy level and the probability that the electron will occupy these states (Fermi function). Therefore, the number of electrons and holes found between the energy levels E and $E + dE$ in the conduction band is given by

$$n(E) = f(E)g_c(E)dE \tag{2.8}$$
$$p(E) = (1 - f(E))g_c(E)dE$$
$$\tag{2.9}$$

The number of electrons and holes found in the valence band between the energy levels E and $E + dE$ is given by

$$n(E) = f(E)g_v(E)dE \tag{2.10}$$
$$p(E) = (1 - f(E))g_v(E)dE$$
$$\tag{2.11}$$

Here $n(E)$ and $p(E)$, are the electron and hole concentration per unit volume per unit energy. When we consider charge transport in the semiconductor, we

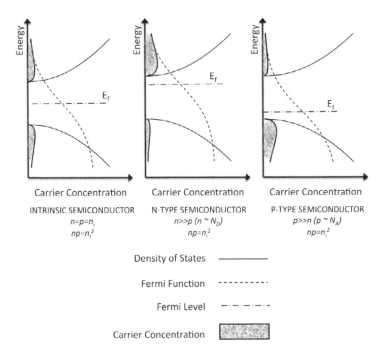

FIGURE 2.4
The free carrier density for an intrinsic, n-type doped and p-type doped semi-conductor.

will be mainly interested in the number of electrons in the conduction band, and the number of holes in the valence band.

In the band-gap, the value of the Fermi function is not zero for temperatures greater than 0K. Yet, the density of states is zero. This implies that although the probability of finding carriers is non-zero, there are no states for the carriers to occupy. Hence, the total number of carriers in the band-gap is zero. Once there are defects in the semiconductor, states begin to appear in the band-gap and with it the population of carriers in the gap increases. The qualitative nature of the Fermi function at different temperatures is shown in Fig. 2.3.

Next, we ask the question — what is the total number of free electrons per unit volume in the conduction band at a given temperature? The answer is found by summing the electrons at all energy levels above E_c, i.e.,

$$n = \int_{E_c}^{\infty} f(E)g_c(E)dE \tag{2.12}$$

$$= 2\left(\frac{2\pi m_e^* kT}{h^2}\right)^{3/2} e^{\frac{-(E_c-E_f)}{kT}}$$

$$= N_c e^{\frac{-(E_c-E_f)}{kT}}$$

From the above set of equations, we define N_c, the *effective density of states* in the conduction band, which describes the overall effect of the energy distribution of states in determining the total population of electrons in the conduction band.

Equivalently, we can define the effective density of valence band states, N_v, to aid in determining the total number of holes per unit volume in the valence band.

$$p = \int_{E_v}^{\infty} (1 - f(E))g_v(E)dE \tag{2.13}$$

$$= 2\left(\frac{2\pi m_h^* kT}{h^2}\right)^{3/2} e^{\frac{-(E_f-E_v)}{kT}}$$

$$= N_v e^{\frac{-(E_f-E_v)}{kT}}$$

With the knowledge of the effective density of conduction band, N_c, effective density of valence band states, N_v, the position of the conduction band edge, E_c, the position of the valence band edge, E_v, and the Fermi level, E_f, the free electron and hole concentrations can be completely determined.

2.3.5 Doping

When the semiconductor is pure, it is known as being *intrinsic* and the semiconductor has an equal number of electrons and holes since it is charge neutral. This concentration of carriers in an intrinsic semiconductor at thermal equilibrium is called the *intrinsic carrier concentration* denoted by n_i. Thus, $n = p = n_i$ where n is the free electron concentration per unit volume, and p is the hole concentration per unit volume. In an intrinsic semiconductor, the position of the Fermi level is called the *intrinsic Fermi level*, E_i. Thus, according to Equation 2.12 and Equation 2.13, for an intrinsic semiconductor

$$n = n_i = N_c e^{\frac{-(E_c-E_i)}{kT}} \tag{2.14}$$

$$p = n_i = N_v e^{\frac{-(E_i-E_v)}{kT}}$$

In order to manipulate the concentration of carriers in the semiconductor, we can manipulate the position of the Fermi level. This is achieved by doping the intrinsic semiconductor with foreign atoms. When a semiconductor is doped, the foreign atoms force themselves into the bonding matrix of the semiconductor. If the dopant atom has more electrons in the valence shell as compared to the semiconductor atom, the bonding results in some unbonded electrons that are free to conduct. This form of doping, which leads to the semiconductor having excess electrons, is called *n-type doping*. Equivalently, if the semiconductor is doped with atoms having less electrons in the valence shell as compared to the semiconductor atoms, it results in the presence of excess vacancies or holes. Such a doping is called *p-type doping*. For example, if we consider the crystalline silicon semiconductor, silicon atoms have 4 electrons in their valence shell. Doping the semiconductor with Phosphorous atoms that have 5 electrons in the valence shell, leads to n-type doping. On the other hand, doping the silicon semiconductor with Boron, which has 3 electrons in the valence shell leads to p-type doping.

When the semiconductor is doped n-type, the Fermi level moves closer to the conduction band edge as there are more electrons in the semiconductor. Increasing the dopant concentration pushes the Fermi level closer to the band edge. If the semiconductor is doped p-type the Fermi level moves closer to the valence band edge as there are more holes in the semiconductor. According to Equation 2.12 and Equation 2.13, after doping, the carrier concentration called the *extrinsic carrier concentration* is

$$n = N_c e^{\frac{-(E_c - E_f)}{kT}} \tag{2.15}$$

$$p = N_v e^{\frac{-(E_f - E_v)}{kT}}$$

Note the Fermi level position is E_f and different from E_i. Therefore, n and p are related to n_i as

$$n = n_i e^{\frac{E_f - E_i}{kT}} \tag{2.16}$$

$$p = n_i e^{\frac{E_i - E_f}{kT}}$$

$$\tag{2.17}$$

In a n-type semiconductor the Fermi level is above the intrinsic Fermi level, and hence $n \gg p$. On the other hand, in the p-type semiconductor, the Fermi level is below the intrinsic Fermi level and $p \gg n$. Figure 2.4 illustrates the carrier concentration for intrinsic, n-type doped and p-type doped semiconductors. The dopant concentration for the n-type semiconductor is called the *donor concentration* quantified by the number of donor ions per unit volume, N_d. Since the donor ions have donated electron, they are positively charged. The dopant concentration for the p-type semiconductor is called the *acceptor concentration* quantified by the number of acceptor ions per unit volume, N_a. Since the acceptor ions have accepted electrons (and donated holes), they are negatively charged.

If the doping in the semiconductor is very strong, the Fermi level moves very close to the band edges. In the case of very strong n-type doping if

$E_c - E_f$ may become comparable to kT and the Boltzmann approximation of the Fermi function may become invalid. Similarly, in the case of very strong p-type doping $E_f - E_v$ may become comparable to kT and once again the Boltzmann approximation for $1 - f(E)$ cannot be used. Such semiconductors that are strongly doped (dopant concentration $\geq N_c, N_v$) are called *degenerate semiconductors*.

2.3.6 Mass-Action Law

If we consider a semiconductor in thermal equilibrium, free charge carriers are continously created due to the presence of thermal energy. The generation rate of free charge carriers is $G = g(T)$, and is purely a function of the temperature and the properties of the semiconductor crystal lattice. The generation of electron and holes in the semiconductor is balanced by their recombination and this balance of generation and recombination maintains thermal equilibrium in the semiconductor. The recombination is proportional to $R = npr(T)$, where $r(T)$ is again purely a function of temperature and crystal property. Equating the generation and recombination rates,

$$np = g(T)/r(T) \tag{2.18}$$

Since np is a function of the temperature and material property alone, it is a constant at a fixed temperature for a given material. For an intrinsic semiconductor, $n = p = n_i$, and therefore the constant is n_i^2. Note that for both intrinsic and doped (extrinsic) semiconductors,

$$np = n_i^2 = N_c N_v e^{\frac{-E_g}{kT}} \tag{2.19}$$

where E_g is the energy gap. This is called the *Mass-Action law*.

Therefore, for n-type semiconductors, $N_d >> n_i$, and $n \approx N_d$ and due to Equation 2.19 $p \approx n_i^2/N_d$. For p-type semiconductors, $N_a >> n_i$ and $p \approx N_a$ and due to Equation 2.19 $n \approx n_i^2/N_a$.

2.3.7 Charge Transport

2.3.7.1 Drift Current

In the absence of an electric field the charge carriers move about with random velocities energized by thermal energy. The mean square thermal velocity of the electron is given by $(3kT/m_e^*)^{1/2}$. Due to the completely random motion of electrons, there is no net current. The electrons collide with themselves and the crystal and the average time interval between collisions is called the *mean scattering time for electrons* denoted by τ_{me}.

When a voltage is applied across a block of semiconductor, an electric field is established with the semiconductor. The electrons therefore experience an acceleration in the direction of the field in between the scattering. The force on the electron due to applied field is $-q\xi$. The effective momentum transferred to the electrons is thus $-q\xi\tau_{me}$. If we define the drift velocity of the electrons v_d, the momentum of the electron is $m_e^* v_d$. Thus, we can establish a linear relation between electric field and drift velocity as

$$-q\xi\tau_{me} = m_e^* v_d \qquad (2.20)$$

The drift velocity is proportional to the applied electric field and the constant of proportionality is called the *electron mobility* defined as $\mu_e = q\tau_{me}/m_e^*$.

The applied electric field not only causes a drift in electrons but also a drift in the holes in the opposite direction. The hole drift is characterized by the hole mobility defined as $\mu_h = q\tau_{mh}/m_h^*$. This drift of electrons and holes results in a current, and the current density is the sum of the electron and hole current densities. The total drift current density is given by

$$J_{dr} = J_{dr_e} + J_{dr_h} = q(\mu_e n + \mu_h p)\xi \qquad (2.21)$$

where n and p are the carrier concentrations of electrons and holes, respectively. The conductivity of the semiconductor is thus defined from the above relation to be $q(\mu_e n + \mu_h p)$. In the case of doped semiconductors one type of carrier is much larger than the other type — for example in n-type semiconductors, $n >> p$ and $J_{dr} \approx J_{dr_e}$.

2.3.7.2 Diffusion Current

Another important means of charge transport in semiconductors is due to diffusion. Spatial variation in carrier density within the material leads to charge transport from the regions of higher concentration to the regions of lower concentration. The total diffusion current due to electrons and holes is given by

$$J_{df} = J_{df_e} + J_{df_h} = qD_e dn/dx - qD_h dp/dx \qquad (2.22)$$

The diffusion current along the x-direction depends on the gradient of carrier concentration in this direction denoted by the first differentials dn/dx and dp/dx. D_e and D_h are the *diffusion constants* for electrons and holes, respectively. The diffusion constants are related to the mobility of carriers by the Einstein relations

$$D_e = kT\mu_e/q \qquad (2.23)$$
$$D_h = kT\mu_h/q$$

$$(2.24)$$

The total current in the semiconductor is the sum of the drift and diffusion components of electrons and holes.

Einstein's Relation

Einstein's relations relate the diffusion coefficient to the mobility of carriers at thermal equilibrium. These relations are not only applicable to electronic charges, but to a wide variety of systems that have drift and diffusion based transport, and follow the Boltzmann statistics.

Let us consider the semiconductor with only one type of carriers (say electrons). At thermal equilibrium, the net current (drift and diffusion) in the semiconductor must be zero. Therefore,

$$Ddn/dx - q\mu n\xi = 0 \qquad (2.25)$$

The electric field ξ can be written in terms of the potential gradient, φ, as $\xi = -d\varphi/dx$. According to the Boltzmann distribution of carriers, $n = Ce^{-q\varphi/kT}$ (C is a constant) and $dn/dx = -nq/kT(d\varphi/dx)$. Using these relations in the above equation we obtain Einstein's relation

$$D = kT\mu/q \qquad (2.26)$$

2.4   Semiconductor Junctions

Next, we consider a junction formed between a semiconductor and another material that may be a semiconductor, metal, or insulator. If we label the semiconductor as A and the second material as B, we can define the Fermi level positions in the two materials to be at E_{fA} and E_{fB}. The Fermi functions in the two materials are given as

$$f_A(E) = \frac{1}{1 + e^{\frac{E - E_{fA}}{kT}}} \qquad (2.27)$$

$$f_B(E) = \frac{1}{1 + e^{\frac{E - E_{fB}}{kT}}}$$

$$(2.28)$$

If $g_A(E)$ and $g_B(E)$ are the density of states in materials A and B, we can estimate the electron concentration in these states, n_A, and n_B, as

$$n_A(E) = f_A(E)g_A(E) \tag{2.29}$$
$$n_B(E) = f_B(E)g_B(E)$$
$$\tag{2.30}$$

Equivalently, the hole concentration in materials A and B is given by

$$p_A(E) = (1 - f_A(E))g_A(E) \tag{2.31}$$
$$p_B(E) = (1 - f_B(E))g_B(E)$$
$$\tag{2.32}$$

When the two materials are joined to form a junction, the electrons from each material will attempt a flow depending on the concentration gradient such that the flow rates are balanced. This is the demand of thermal equilibrium. The flow rate from material A to material B is proportional to the number of electrons in material A and the number of empty states in material B and is given by $n_A p_B$. On the other hand the flow rate from material B to material A is given by $n_B p_A$. Equating the two flow rates.

$$n_A p_B = n_B p_A \tag{2.33}$$
$$\Rightarrow \quad f_A g_A (1 - f_B)g_B = f_B g_B (1 - f_A)g_A$$
$$\Rightarrow \quad f_A = f_B$$
$$\Rightarrow \quad E_{fA} = E_{fB}$$

Thus, after junction formation, the Fermi levels of both materials align under the conditions of thermal equilibrium. Using this fact, we can draw *band diagrams* for semiconductor junctions and analyze the properties of these junctions.

The alignment of the Fermi level can also be intuitively understood using the "bucket of water" analogy. If we bring two buckets with different levels of water and form a junction removing the wall between them, the water will flow from the higher level to the lower level such that, at equilibrium, both buckets have the same water level.

2.5 Metal-Semiconductor Junction

When a metal and a semiconductor are joined, two possible types of contact can result, depending on the combination of metal and semiconductor used. The metal is defined by the position of the Fermi level, which in turn determines its work function, $q\Phi_M$, which is the energy difference between the

Fermi level and vacuum. The semiconductor is defined by its work function, $q\Phi_S$, its energy gap, E_g, and its electron affinity, $q\chi_S$, which is the energy difference between the conduction band edge and vacuum. Depending on the semiconductor and metal used, the contact may be *rectifying* or *ohmic*. A rectifying contact allows current to pass in one direction only while it blocks the movement of carriers in the opposite direction. On the other hand, an ohmic contact allows current to pass in either direction.

If the semiconductor is n-type with $\Phi_S < \Phi_M$ or if it is p-type with $\Phi_S > \Phi_M$ the contact formed is rectifying or blocking. On the other hand, if $\Phi_S > \Phi_M$ with an n-type semiconductor or $\Phi_S < \Phi_M$ in a p-type semiconductor, the contact is ohmic.

2.5.1 Rectifying contact

Let us first consider the formation of a rectifying contact with an n-type semiconductor under the condition that $\Phi_S < \Phi_M$. The band diagrams of the two materials are shown in the Figure 2.5. When the junction is formed by bringing the semiconductor and metal together, the electrons in the semiconductor rush to the metal (since the Fermi level in the semiconductor is higher than that of the metal), till the Fermi levels align.

The flow of electrons into the metal from the semiconductor on contact causes a negative charging of the metal, thereby repelling further flow of electrons from the conduction band of the semiconductor into the metal. The Fermi level of the metal is not affected because there are many more electrons in the metal than there are free electrons in the semiconductor before contact and the small flow of electrons from the semiconductor does not add much. However, since the semiconductor at the semiconductor-metal interface has lost a lot of electrons, the Fermi level in that region of the semiconductor moves away from the conduction band edge and closer to the valence band edge. As we move away from the interface into the semiconductor (say the x direction), the amount by which the Fermi level has shifted away from the conduction band edge varies with the distance. Beyond some distance away from the semiconductor-metal interface, the Fermi level has not shifted at all since this region of the semiconductor did not have to donate charge to the metal. We draw this situation by keeping the Fermi level flat and distorting the conduction and valence band edges as a function of the distance. This implies that the conduction and valence bands of the semiconductor bend as shown in Figure 2.5. This deformation of energy bands is something termed as *band bending*, and diagrams that model the physics of such interfaces involving band bending are called *band bending diagrams*. As we move further away from the junction into the semiconductor, there has been no transfer of charge and hence the bands remain flat. The semiconductor near the junction has no conduction electrons in it and is depleted of electrons in the conduction band. This region is therefore called the *depletion region*, or the *space charge region*. The term space charge region is due to the fact that this region of the semicon-

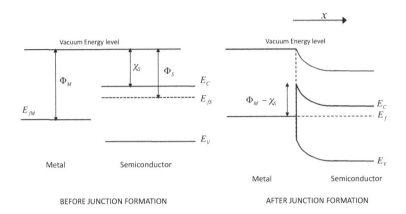

FIGURE 2.5
The formation of a rectifying metal-semiconductor junction.

ductor is filled with fixed dopant ion (in this case the positively charged donor ions) creating a space filled with immovable charges. This depletion layer acts like a potential barrier with *barrier height* $\Phi_b = \Phi_M - \chi_S$ for electrons moving from the metal to the semiconductor. The bending of bands with distance implies a potential gradient varying with distance, $\varphi(x)$, and hence the presence of an electric field, $\xi(x) = -d\varphi(x)/dx$ in the semiconductor near the junction.

Fermi Level Pinning

The intuitive understanding of the formation of the barrier and the definition of the barrier height as $\Phi_b = \Phi_M - \chi_S$ is based on the Schottky-Mott model. However, experiments have shown that the properties of the surface at the interface plays a significant role in determining the barrier height, much more than the work function of the metal itself. Due to the presence of states at the interface, the free movement of the Fermi level appears to be hampered and the Fermi level appears to be "pinned". Recent studies have shown that Fermi level pinning is a direct consequence of the bonding at the interface.

2.5.1.1   Analysis of the Electrostatics at Junctions

In order to quantify the nature of the depletion region, we perform an electrostatic analysis of the semiconductor near the junction. The starting point of such an analysis is the *Poisson's Equation*, which is the differential form of Gauss' Law (introduced in the previous chapter). The Poisson's equation relates the electric field in the semiconductor to the charge per unit volume

$$\frac{d\xi}{dx} = \frac{\rho}{\epsilon_s} \tag{2.34}$$

where ξ is the electric field, ρ the charge density per unit volume in the region under study, and ϵ_s the permittivity of the region under study. Since the electric field is the gradient of the potential, φ, the Poisson's equation also relates the potential to the charge density as

$$\frac{d^2\varphi}{dx^2} = -\frac{\rho}{\epsilon_s} \tag{2.35}$$

We now use the Poisson equation to analyze the depletion region of the semi-conductor when a rectifying metal-semiconductor contact is present.

Since the depletion region is depleted of free carriers, the charge in the depletion region is due to the presence of fixed donor dopant ions which have a concentration per unit volume of N_d. Thus, Equation 2.35 becomes

$$\frac{d^2\varphi}{dx^2} = -\frac{qN_d}{\epsilon_S} \tag{2.36}$$

where φ is the potential, x is the spatial distance in the direction perpendicular to the plane of the junction, and ϵ_S is the permittivity of the semiconductor. Solving for the potential, we first estimate the electric field ξ as,

$$\xi = -\frac{d\varphi}{dx} = \frac{qN_d x}{\epsilon_S} + const. \tag{2.37}$$

The constant of integration can be obtained by noting that the electric field far away from the junction in the semiconductor is zero since the bands are flat. Hence, if x_d is the width of the depletion region, $d\varphi/dx = 0$ at $x = x_d$, implies

$$\frac{d\varphi}{dx} = -\frac{qN_d(x - x_d)}{\epsilon_S} \tag{2.38}$$

Integrating further,

$$\varphi(x) = -\frac{qN_d(\frac{x^2}{2} - x_d x)}{\epsilon_S} + const. \tag{2.39}$$

At $x = 0$, $\varphi = const.$, and at $x = x_d$, $\varphi = \frac{qN_d x_d^2}{2\epsilon_S} + const.$ Thus, the potential difference due to the band bending in the semiconductor — which is the barrier (Φ_i) seen by the carriers moving from the semiconductor to the metal — is given by $\Phi_i = \varphi(x_d) - \varphi_0 = \Phi_M - \Phi_S$. From Equation 2.39 it is seen that

$$\Phi_i = \frac{qN_d x_d^2}{2\epsilon_S} \tag{2.40}$$

The width of the depletion region is then easily found to be $(2\epsilon_S\Phi_i/qN_d)^{1/2}$.

2.5.1.2   Currents across the Junction

At any given temperature, there exists drift and diffusion currents across the barrier. The current from the metal to the semiconductor is equaled by the current from the semiconductor to the metal thereby defining the thermal equilibrium condition. The drift current component is very small. It exists because of the electron–hole pairs formed in the depletion region due to thermal excitations. These thermally generated carriers are propelled by the electric field in the region with the electrons being driven towards the semiconductor, and the holes being driven towards the metal. This current is independent of the barrier height. A much larger component of current is due to diffusion. The diffusion current component exists because some electrons in the metal are thermally excited to overcome the potential barrier and diffuse into the semiconductor conduction band while some electrons in the conduction band of the semiconductor have enough thermal energy to diffuse from the semiconductor into the metal. In thermal equilibrium these currents are equal and cancel each other.

When a voltage is applied across the junction, the system moves out of thermal equilibrium and the Fermi level of the semiconductor and metal will no longer remain aligned. If the metal is kept at a lower potential compared to the semiconductor by an externally applied voltage V_a, there will be a greater negative charge on the metal and the Fermi level of the metal rises. This situation is called *reverse bias*. The barrier seen by the electrons in the metal remains at Φ_b. However, as the Fermi level of the metal climbs up with the applied voltage, the bands in the semiconductor are further deformed, and the barrier seen by the electrons in the n-type semiconductor is increased to $\Phi_i + V_a$. This results in the diffusion current becoming negligible since the potential barrier is too large. The contribution from the drift current constitutes a portion of the current flow as a small current of generated carriers drift across the depletion region. When the metal is instead kept at a higher potential with respect to the semiconductor, the negative charge on the metal is reduced and the Fermi level in the metal falls. This situation is called *forward bias*. This lowering of the metal Fermi level, reduces the deformation of the bands in the semiconductor. Thus, the barrier for the flow of electrons from the semiconductor to the metal is reduced to $\Phi_i - V_a$. The increased diffusion of electrons from the semiconductor to the metal causes a significant flow of current across the junction. Therefore, the current in forward bias is much larger as compared to the current in reverse bias. The metal-semiconductor junction with rectifying nature is also called a *Schottky diode*, and effectively permits current in only one direction i.e., from the metal to the semiconductor, while significantly blocking current in the reverse direction. This is the rectifying nature of the junction. The current through the junction is defined as,

$$I = AT^2 e^{-\frac{q\Phi_b}{kT}} \frac{qm_e^* k^2}{2\pi^2 h^3} \left(e^{\frac{qV_a}{kT}} - 1 \right) \tag{2.41}$$
$$= I_s \left(e^{\frac{qV_a}{kT}} - 1 \right)$$

Here, A is the cross-section area of the junction, T is the temperature, k the Boltzmann's constant, h the Planck's constant, and the term $\frac{qm_e^* k^2}{2\pi^2 h^3}$ is the Richardson's constant.

2.5.1.3 Impedance of the Schottky diode

The rectifying metal-semiconductor junction can be thought of as a resistor of resistance, R_s, and capacitor of capacitance, C_s, connected in parallel.

The capacitive component of the junction arises due to the presence of the depletion layer in the semiconductor and near the interface. It was seen earlier that the width of the depletion layer at thermal equilibrium is given by $x_d = (2\epsilon_S \Phi_i/(qN_d))^{1/2}$. When a voltage bias V_a is applied across the junction, the depletion layer thickness is modified to be $x_d = (2\epsilon_S(\Phi_i - V_a)/(qN_d))^{1/2}$. The capacitance of the rectifying contact therefore is,

$$C_s = \epsilon_S A/x_d = A \left(\frac{q\epsilon_S N_d}{2(\Phi_i - V_a)} \right)^{1/2} \tag{2.42}$$

Here A is the cross-sectional area of the junction. Another method of calculating this capacitance is to study the change of charge in the depletion region with a change of applied voltage, i.e. $C_s = dQ/dV_a = qN_d A(dx_d/dV_a)$. This capacitance component arising due to the depletion region is called the *depletion capacitance*.

The resistance of the junction is $R_s = dV_a/dI$ where I is the current through the junction. From Equation 2.41 it is seen that

$$R_s = \frac{kT}{qI_s e^{\frac{qV_a}{kT}}} \approx \frac{kT}{qI} \tag{2.43}$$

2.5.1.4 Rectifying Contact with p-type Semiconductor

A rectifying contact can also be formed between a metal and p-type semiconductor if $\Phi_S > \Phi_M$. The band-bending diagram is symmetrical to a metal and n-type semiconductor junction with the barrier forming at the valence band edge thereby blocking the movement of holes (which are the majority carriers).

2.5.2 Ohmic Contact

Next, we consider the formation of an ohmic contact with an n-type semiconductor. An ohmic contact forms under the condition that $\Phi_S > \Phi_M$.

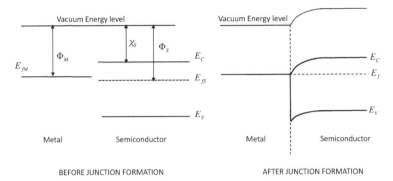

FIGURE 2.6
The formation of an ohmic metal-semiconductor junction. The semiconductor is n-type.

When the junction between the semiconductor and metal is formed in such a case, the electrons from the metal diffuse into the semiconductor thereby leading to the bands bending downward. Figure 2.6 illustrates the formation of the ohmic metal-semiconductor junction with an n-type semiconductor. Such a contact allows for charge transport in both directions with the current being almost linearly proportional to the voltage applied. Such a contact is called an *ohmic contact.*

A p-type semiconductor with $\Phi_S < \Phi_M$ also forms an ohmic contact.

However, since the cleaved face of the semiconductor is filled with dangling bonds, any junction formed along this face with a metal is not likely to allow the free movement of the Fermi level since there are a large number of carrier capturing states at this interface. In practice, it is not straightforward to obtain an ohmic contact with a metal and semiconductor by simply choosing their work functions. A more practical approach is to heavliy dope the semiconductor interface thereby leading to a barrier contact and very sharp band bending in the semiconductor that effectively reduces the barrier width. Since the barrier width is low, the carriers simply tunnel through the barrier thereby creating a good contact.

2.6   p-n Junction

In the previous section we looked at the case when a metal is brought in contact with a semiconductor thereby forming a metal-semiconductor junction. Nex,t we study the junction formed between a p-type semiconductor and an n-type semiconductor. This junction is called a *p-n junction.* When the semi-

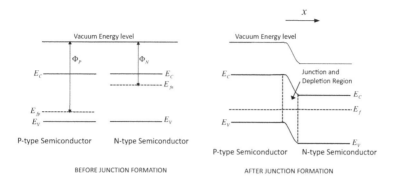

FIGURE 2.7
The formation of a p-n junction.

conductor is the same, the Fermi level of p-type semiconductor is close to the valence band, while the Fermi level of the n-type semiconductor is close to the conduction band. In other words the work function of the n-type semiconductor, Φ_{Sn}, is less than the work function of the p-type semiconductor Φ_{Sp}.

Since $\Phi_{Sn} < \Phi_{Sp}$, there has to be a net flow of electrons from the n-side to the p-side and a net flow of holes from the p-side to the n-side due to diffusion. Comparing this case with the metal semiconductor junction we studied before, the contact formed by with a p-type and n-type semiconductor is always rectifying with current being permitted to flow only from the p-side to the n-side and not the other way round. After the formation of the junction, the Fermi level of the two semiconductors level to equilibrium, and the bands are deformed in both the p-type semiconductor and n-type semiconductor forming a depletion layer at the interface.

2.6.1 Analysis of p-n Junction

The analysis of the electrostatics of the p-n junction is done using Poisson's equation similar to the case of the metal-semiconductor rectifying junction. Writing Poisson's equation for the n- and p-sides

$$\frac{d^2\varphi_n}{dx^2} = -\frac{qN_d}{\epsilon_S}$$

$$\frac{d^2\varphi_p}{dx^2} = \frac{qN_a}{\epsilon_S}$$

(2.44)

where φ_n and φ_p are the potentials for the n- and p- sides, respectively, x the spatial distance, and ϵ_S is the permittivity of the semiconductor. The electric fields in the n- and p-side of the junction are given by

$$\xi_n = -\frac{d\varphi_n}{dx} = \frac{qN_d(x - x_{n-d})}{\epsilon_S} \tag{2.45}$$

$$\xi_p = -\frac{d\varphi_p}{dx} = -\frac{qN_a(x - x_{p-d})}{\epsilon_S}$$

where x_{n-d} and x_{p-d} are the widths of the depletion layer in the n- and p-side, respectively. The charge balance across the junction demands

$$N_a x_{d-p} = N_d x_{d-n} \tag{2.46}$$

The potential profile in the n- and p-type semiconductors is found to be

$$\varphi_n = -\frac{qN_d(\frac{x^2}{2} - x_{n-d}x)}{\epsilon_S} + const \tag{2.47}$$

$$\varphi_p = \frac{qN_a(\frac{x^2}{2} - x_{p-d}x)}{\epsilon_S} + const$$

Thus, the potential across the barrier is the potential across, $x = x_{n-d}$ and $x = x_{p-d}$ and is given by

$$\Delta\varphi_{p-n} = q\frac{N_a x_{p-d}^2 + N_d x_{n-d}^2}{2\epsilon_S} \tag{2.48}$$

2.6.2 Currents across the Junction

In equilibrium, the net drift and diffusion currents are balanced out. Just as in the case of a metal-semiconductor rectifying junction, there appears a potential barrier for electrons to diffuse from the n-type semiconductor into the p-type semiconductor, and also for holes to diffuse from the p-type semiconductor into the n-type semiconductor. In addition to the diffusion current, there appears a drift current due to electron–hole pairs created due to thermal excitation in the depletion region. These carriers are accelerated by the in-built electric field with the holes pushed to the p-side and electrons to the n-side. At equilibrium, the total current across the junction is the same in both directions, and the net current is zero.

Suppose a voltage is applied such that the n region is maintained at a higher potential compared to the p region, the bands are deformed further, creating larger potential barriers for both electrons and holes to move across the junction along with increasing the width of the depletion layer. In this situation of reverse bias, the only current component is due to the drift of minority carriers.

On the other hand, if the p region is maintained at a higher potential compared to the n-region, the band bending decreases thereby lowering the potential barrier, and reducing the width of the depletion region. The diffusion current increases exponentially thereby increasing current flow.

Therefore a p-n junction diode has a rectifying nature where it allows current in one direction - p side to n side , while blocking the current in the opposite direction - i.e. n side to p-side.

2.7 Transistors

The primary idea behind the transistor is to have a resistor whose resistance is controlled electrically. There are several types of transistor architectures with the most popular being bipolar junction transistors (BJTs), junction field effect transistors (JFETs), and metal oxide semiconductor field effect transistors (MOSFETs). In our discussions below and the remainder of this book we pay special attention to the MOSFET.

The primary idea behind the MOSFET is that the resistance and hence the current between two terminals called the *drain* and *source* is controlled electronically. The electronic control is achieved by controlling the free carrier concentration between the drain and source by modulating the electric field using another electrode called the *gate*. A crystalline silicon MOSFET also has an electrode connected to the body or bulk semiconductor of the MOSFET. This terminal is called the *body*. By controlling the voltage on the drain, source and gate electrodes, we bias the MOSFET in the region of operation we desire. The subscipts d or D, s or S, g or G, and b or B are used with voltages or currents to identify the electrical conditions of the drain, source, gate and body electrodes, respectively. The following sections discuss the details of MOSFET operation. These ideas easily translate to field effect devices based on non-crystalline semiconductors.

2.7.1 MOS Capacitor

In this section we study the basic ideas related to field effect based semiconductor devices. The starting point of our study of field effect devices is the metal-oxide-semiconductor (MOS) capacitor shown in Figure 2.8. Though we specify "oxide", we imply the use of an insulator and hence the terminology might as well be the metal-insulator-semiconductor (MIS) capacitor. Nevertheless, we use the term MOS irrespective of the materials used. In this discussion the semiconductor is crystalline silicon.

2.7.1.1 Structure

The MOS capacitor consists of a stack of metal(top metal)-insulator-semiconductor layer with the semiconductor having a back metal ohmic con-

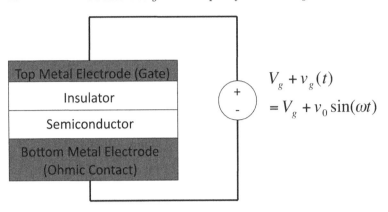

FIGURE 2.8

Metal-oxide-semiconductor (MOS) capacitor. The term "oxide" is due to the typically used silicon oxide as a dielectric in conventional crystalline silicon MOS transistors. However, any insulator can be present instead of the oxide and the device is equivalently termed metal-insulator-semiconductor.

tact. The back metal contact is kept at ground potential throughout this discussion and acts a source and sink of carriers for the semiconductor. If the semiconductor is p-doped we call the MOS capacitor an NMOS capacitor and if the body is doped n-type we call it a PMOS capacitor. The reasons for these labels has to do with the type of the minority carriers and will soon become apparent. For the sake of this discussion, we assume that the semiconductor is p-type and hence we study the NMOS capacitor. The top metal contact is called the *gate* and any charge on the gate electrode results in an electric field in the vertical direction, which in turn results in the modulation of carriers at the semiconductor-insulator interface. This fundamental notion of controlling the surface charge at the semiconductor-insulator interface via the application of a voltage on the gate with respect to the back metal electrode, is the principle of operation of the *field effect transistor*. The gate-ground voltage is given the symbol V_g.

Once the MOS structure is formed, the Fermi levels in the gate metal, insulator, and semiconductor align, and the semiconductor may experience some band bending because of the work function difference between the semiconductor and metal. In order to remove the bandbending in the semiconductor, the gate must be kept at some potential, $V_g = V_{fb}$, called the *flatband potential*.

2.7.1.2   Operation

When V_g is lowered significantly below V_{fb}, the gate electrode is negative with respect to ground, and holes in the semiconductor accumulate at the semiconductor-insulator interface. This is called the *accumulation mode* of operation.

As V_g is increased a little past V_{fb}, the holes are driven away from the interface, and the charge on the gate is compensated for by the negatively charged fixed dopant acceptor ions in the semiconductor. If the acceptor ion concentration is N_A per unit volume, the band bending in the semiconductor depletes the semiconductor-insulator interface of any free carriers to a distance x_d. The only free carriers present in this depletion region is due to the thermal generation of electron-hole pairs within the region. This is called the *depletion mode* of operation. The band bending in the semiconductor can be analyzed by solving Poisson's equation.

As V_g is increased further, the band bending at the semiconductor-insulator interface becomes so steep (implying the electric field is so strong) that the electrons begin to accumulate at the surface. Since the channel now contains carriers of the opposite type as compared to the accumulation mode of operation, this is called the *inversion mode* of operation. The reason why the p-type semiconductor based MOS capacitor is called the NMOS capacitor is primarily due to the type of carriers involved in the formation of the inversion layer. The inversion layer is also called the *channel* of the MOSFET. Thus, the NMOS FET is also called an *n-channel MOSFET* and a PMOS FET is also called a *p-channel MOSFET*.

2.7.1.3 Analysis

Consider a MOS capacitor structure formed of a p-type semiconductor. Before the formation of the MOS structure we define $\varphi_i = (E_f - E_i)/q$ where E_i is the intrinsic Fermi level position, and E_f is the Fermi level position of the doped semiconductor at thermal equilibrium. Thus, φ_i is indicative of the degree of doping and hence the hole concentration in the semiconductor.

After the formation of the MOS structure, there is an exchange of carriers, and the Fermi levels align. We now consider the MOS capacitor in depletion mode operation as shown in Figure 2.9. The position of intrisic Fermi level, E_i, is now a function of x, while E_f remains flat. Poisson's equation for this situation is

$$\frac{\partial^2 \varphi}{\partial x^2} = \frac{qN_a}{\epsilon_S} \tag{2.49}$$

Here $\varphi = (E_f - E_i(x))/q$ is the potential which is a function of x. ϵ_S is the permittivity of the semiconductor. The value of φ at $x = 0$ is $\varphi_{sf} = (E_f - E_i(0))/q$ and $\varphi_{sf} - \varphi_i$ is called the *surface potential*. The solution to the above equation defines the width of the depletion layer to be

$$x_d = \left(\frac{2\epsilon_S |\varphi_{sf} - \varphi_i|}{qN_a} \right)^{1/2} \tag{2.50}$$

The capacitance per unit area of the depletion layer is therefore $C_d = \epsilon_S/x_d = (\epsilon_S qN_a/2|\varphi_{sf} - \varphi_i|)^{1/2}$. The total charge per unit area in the depletion layer is $Q_d = -qN_a x_d = -(2qN_a\epsilon_S|\varphi_{sf} - \varphi_i|)^{1/2}$.

Since the applied gate voltage divides itself between the voltage drop

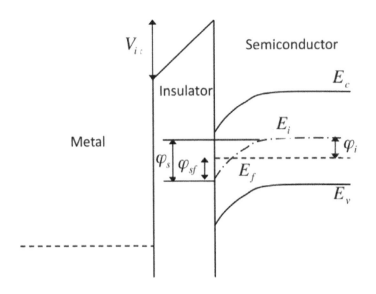

FIGURE 2.9
Band bending at inversion in a MOS capacitor.

across the insulator and the potential at the surface of semiconductor at the semiconductor-insulator interface,

$$\varphi_{sf} - \varphi_i = V_g - V_i \tag{2.51}$$

The charge balance tells us that

$$C_{ox}V_i = Q_f + Q_d \tag{2.52}$$

Here, $C_{ox} = \epsilon_{ox}/t_{ox}$ is the gate capacitance per unit area with ϵ_{ox} being the permittivity of the insulator, and t_{ox} being the insulator thickness. Q_f is the free electron charge per unit area and Q_d is the depletion charge per unit area. Together $Q_f + Q_d$ represent the total charge per unit area in the semiconductor. Thus,

$$V_g = \frac{Q_f + Q_d}{2} + \varphi_{sf} - \varphi_i \tag{2.53}$$

As the gate voltage is increased further the depeltion region widens and the surface potential varies. The moment the surface potential becomes $-\varphi_i$ we have an onset of inversion (Figure 2.9). Just before the onset of inversion, the depletion region attains its maximum width $x_{d-max} = \left(\frac{4\epsilon_s|\varphi_i|}{qN_a}\right)^{1/2}$. This gate voltage at which inversion begins is called the *threshold voltage*, V_{T0}, of the MOS device. Just before the onset of inversion $Q_f \approx 0$, and $Q_d = 2\left(qN_a\epsilon_s|\varphi_i|\right)^{1/2}$. Therefore,

$$V_{T0} = \frac{Q_d}{C_{ox}} + 2|\varphi_i| + V_{fb} \qquad (2.54)$$

After inversion, the surface of the semiconductor at the insulator interface is populated with electrons (a sheet charge of electrons). Since the electron concentration increases exponentially with any further band bending, the potential at the surface is more or less pinned to the value $-\varphi_i$ after inversion. Moreover, the electric field due to the gate is almost completely screened by the thin layer of electron charge present at the semiconductor-insulator interface. At this point Poisson's equation can be written as

$$\frac{\partial^2 \varphi}{\partial x^2} \approx \frac{qn_i e^{\frac{q\varphi}{kT}}}{\epsilon_S} \qquad (2.55)$$

where n_i is the intrinsic carrier concentration in the semiconductor. In order to solve this equation, we multiply both sides by $2d\varphi/dx$, and note that the left hand side is $2(d\varphi/dx)d^2\varphi/dx^2 = d((d\varphi/dx)^2)/dx = d\xi^2/dx$. Thus we have

$$\frac{d\xi^2}{d\varphi} \approx \frac{2qn_i e^{\frac{q\varphi}{kT}}}{\epsilon_S} \qquad (2.56)$$

resulting in

$$\xi \approx \left(\frac{2kTn_i e^{\frac{q\varphi}{kT}}}{\epsilon_S}\right)^{1/2} \qquad (2.57)$$

2.7.1.4 Capacitance–Voltage Characteristics

In the above discussion of the MOS capacitor we saw that there are three regions of operation — accumulation, depletion, and inversion. We now study the capacitance of the MOS capacitor as the gate voltage is varied, and this measurement is called the *capacitance–voltage (CV) characteristics* of the MOSFET.

In order to measure the MOS capacitance with the gate voltage, we apply a small time-varying signal of a certain frequency, $v_g(t) = v_{g0}sin(\omega t)$ on top of a larger dc voltage which is the gate voltage, V_g. The "large signal" V_g determines the mode of operation of the MOS capacitor. The small fluctuations in this large signal result in small fluctuations in the semiconductor surface charge. By measuring the change in semiconductor charge to the change in applied voltage, we can measure the MOS capacitance.

First we study the CV characterisitcs when ω is low. This implies that the fluctuations in the gate voltage are much slower than the carrier generation time (say for example by thermal generation). In the accumulation mode of opertion, the semiconductor-insulator interface is populated with holes and the surface of the semiconductor is conducting. Therefore, one can think of the MOS capacitor as a parallel plate capacitor with the gate electrode as one plate, and the hole populated conductive semiconductor-insulator interface as the other plate. The capacitance per unit area is thus given by $C_g = C_{ox}$.

In the depletion mode of operation the region near the semiconductor insulator interface is depleted of free carriers. Therefore, we have a situation where the region of the p-type semiconductor beyond the depletion region, i.e, $x > x_d$ acts like a conductive surface. The effective capacitance of the MOS structure can be visualized as the series combination of two capacitors — the gate insulator capacitance of magnitude C_{ox}, and the other, the depletion capacitance of magnitude C_d. The effective capacitance of the MOS capacitor is thus, $C_g = C_{ox}C_d/(C_{ox} + C_d)$.

In the inversion mode of operation, the surface at the semiconductor-insulator interface is populated with electrons thereby making it conductive. Once again, one can think of the MOS capacitor as a parallel plate capacitor with the gate electrode as one plate, and the electron populated conductive semiconductor-insulator interface as the other plate. The MOS capacitance per unit area is again $C_g = C_{ox}$.

The plot of the MOS capacitance versus gate bias for low frequency therefore looks like that shown in Figure 2.10.

What happens when ω becomes large, i.e the fluctuations in the gate voltage are much higher than the rate of carrier generation? When ω becomes very large, there is no significant difference in the measurement of C_g during the accumulation and depletion mode. During accumulation mode, there are a sufficient number of majority carriers that accumulate at the gate-insulator interface. These carriers easily respond to the fluctuations in the gate voltage and the measured MOS capacitance $C_g = C_{ox}$. In the depletion region of operation, the semiconductor-insulator interface is devoid of carriers and once again the measured MOS capacitance is not too different from the low frequency case, i.e. $C_g = C_{ox}C_d/(C_{ox} + C_d)$. However, when the MOS capacitor is biased in the inversion region, the process minority carrier creation via generation-recombination mechanisms is too slow to respond to the change in gate voltage. Therefore, at very high frequency, the measured MOS capacitance remains at $C_g \approx C_{ox}C_d/(C_{ox} + C_d)$. Therefore the capacitance-voltage plot while using high frequency measurements appears as shown in Figure 2.10.

2.7.2   Metal Oxide Semiconductor Field Effect Transistors (MOSFETs)

The above discussion regarding MOS capacitors provide the concepts upon which MOS field effect transistors (MOSFETs) are based. In this section, we discuss the operation of the MOSFET, its current-voltage characteristics, and device parameter extraction.

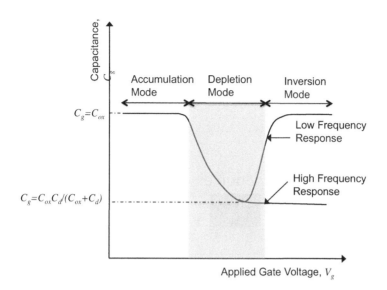

FIGURE 2.10

Capacitance-Voltage (CV) characteristics of a MOS capacitor with a p-type substrate.

2.7.2.1 Structure

The structure of a MOSFET is shown in Figure 2.11 where the source and drain electrodes are added to the MOS capacitor. Both these electrodes must make an ohmic contact with the semiconductor as they are involved in transporting carriers in and out of the semiconductor. The ohmic contact is usually achieved by heavily doping the semiconductor-metal interface n-type for the NMOS transistor and p-type for the PMOS transistor. The distance between these terminals (which also includes the heavily doped region) is the *channel length* of the MOSFET often denoted by L. The width of these terminals is the *channel width* of the MOSFET, and is generally denoted by W. The ratio of the channel width of the MOSFET to the channel length of the MOSFET, W/L, is called the *aspect ratio* of the MOSFET. As discussed earlier, the bulk semiconductor is connected to the body electrode. Another design aspect related to MOSFET design has to do with the gate insulator. For good field effect, the capacitance per unit area of the dielectric, denoted by C_{ox} or C_i, must be large. Since $C_{ox} = \epsilon_{ox}/t_{ox}$, with ϵ_{ox} being the dielectric constant, and t_{ox} being the dielectric thickness, thin insulators with high dielectric coefficients are desirable. However, if the insulator is too thin carriers may tunnel through the insulator from the gate electrode into the semiconductor causing the insulator to break down. This is undesirable. Another important aspect regarding the insulator is that the semiconductor-insulator interface must have

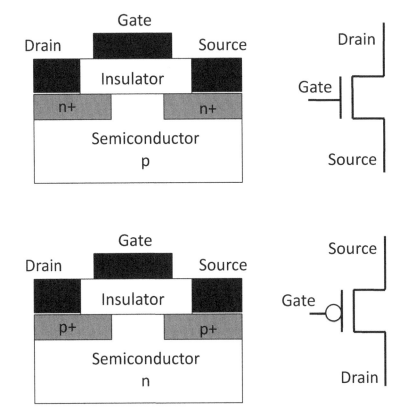

FIGURE 2.11
MOS field effect transistor (MOSFET).

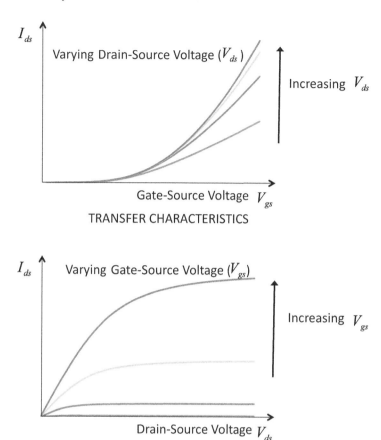

FIGURE 2.12
Transfer and output characteristics of the MOSFET.

minimum "traps". Traps are dangling bonds in the interface and a high trap density can result in loss of carriers thereby reducing the currents in the MOS-FET. Therefore, insulators are so chosen and fabricated that the trap state density at the interface is kept low. For example, silicon oxide forms a good insulator-semiconductor interface with crystalline silicon.

2.7.2.2 Current Voltage Characteristics

With all voltages measured with respect to the source voltage, there are two voltages that control the MOSFET – the drain-source voltage, V_{ds}, and the gate-source voltage, V_{gs}. By controlling V_{gs} and V_{ds} the conductivity of the channel between the drain and source is controlled. This in turn controls the current from the drain to source, I_{ds}. For the present discussion we ignore the influence of the body contact as it will be kept at a constant bias. We assume that the body is grounded for n-channel MOSFETs and is connected to the supply voltage for p-channel MOSFETs.

Let us consider an n-type MOSFET and assume the source electrode (and body electrode) to be grounded. We define the direction along the channel length between drain to source to be the y direction and the direction into the semiconductor (as we saw in the electrostatic analysis with the MOS capacitor) to be the x direction. We also define the MOSFET to have a channel width, W, and channel length L. Typically, the values of V_{gs} and V_{ds} are of the same order of magnitude, while the maximum width of the depletion region and the thickness of the inversion layer is much smaller (several orders of magnitude) as compared to L. Therefore, we can assume that the electric field in the x direction is much larger than the electric field along the y direction.

The primary influence of V_{ds} is that the channel now has a potential that varies from the source to the drain. In other words, the channel potential, $V_{ch}(y)$, varies from 0 to V_{ds} as we move from source to drain. The total charge in the semiconductor per unit area is the sum of the free carrier charge and depletion charge per unit area

$$Q_f + Q_d = C_{ox}(V_{gs} - V_{fb} - 2|\varphi_i| - V_{ch}(y)) \tag{2.58}$$

After inversion, the charge per unit area in the depletion region is given by

$$Q_d(y) = (2q\epsilon_s N_a(2|\varphi_i| + V_{ch}(y)))^{1/2} \approx 2(q\epsilon_s N_a|\varphi_i|)^{1/2} \tag{2.59}$$

Thus, for the elemental section dy along the channel,

$$Q_f(y) = C_{ox}(V_{gs} - V_{fb} - 2|\varphi_i| - V_{ch}(y)) - 2(q\epsilon_s N_a|\varphi_i|)^{1/2} \tag{2.60}$$

We can now define the threshold voltage of the MOSFET in a manner similar to the threshold voltage of the MOS capacitor as

$$V_{T0} = V_{fb} + 2|\varphi_i| + \frac{2(q\epsilon_s N_a|\varphi_i|)^{1/2}}{C_{ox}} \tag{2.61}$$

Thus, Equation 2.60 can be rewritten as,

$$Q_f(y) = C_{ox}(V_{gs} - V_{T0} - V_{ch}(y)) \tag{2.62}$$

Role of the Body Voltage

In the text we have assumed that the body for the n-type MOS-FET is grounded ($V_{bs} = 0$). However, if $V_{bs} \neq 0$, the depletion charge becomes

$$Q_d(y) = (2q\epsilon_s N_a(2|\varphi_i| + V_{ch}(y) - V_{bs}))^{1/2} \tag{2.63}$$

Since, $V_{T0} = V_{fb} + 2|\varphi_i| - Q_d/C_{ox}$, the body-source voltage can be used to modulate the threshold voltage of the transistor.

For a small section of length dy, the voltage drop across the section is $dV = I_{ds}dR$ where dR is the resistance of the section along the y direction. Therefore

$$dV_{ch} = -\frac{I_{ds}dy}{W\mu Q_f(y)} \tag{2.64}$$

where $Q_f(y)$ is the free electron charge density. Note that the negative sign indicates the presence of an electron channel. Equivalently, the negative sign can be placed before Q_f. Substituting for $Q_f(y)$ and solving the above equation results in the drain-source current for the MOSFET to be defined as

$$I_{ds} = \mu C_{ox}\frac{W}{L}\left((V_{gs} - V_{T0})V_{ds} - \frac{V_{ds}^2}{2}\right) \tag{2.65}$$

This equation relating the drain-source current to the bias voltages of the MOSFET is very important and serves us well in this book. However, when $V_{ds} \geq V_{gs} - V_{T0}$ we find, the channel potential near the drain electrode to be so large that number of free carriers are very small. Therefore, the channel charge profile is *pinched off* near the drain electrode. As V_{ds} keeps increasing beyond $V_{gs} - V_{T0}$, this pinch off point travels along the channel length towards the source terminal such that the number of carriers from the pinch-off point to the drain electrode is very small. Since the length of the original channel – from source to pinch-off point – is being modified, this control of the pinch-off point using V_{ds} is called *channel length modulation*. The small number of carriers between the pinch-off point and the drain experience a very large electric field and move very quickly through the semiconductor. The current through the MOSFET is thus limited by the pinched-off region, and remains almost the same for all values of V_{ds} beyond $V_{gs} - V_{T0}$. It is only due to channel length modulation that the current through the MOSFET has a small, and almost linear dependence on V_{ds}. Thus, the MOSFET has two modes of operation above threshold, one with $V_{ds} < V_{gs} - V_{T0}$ called the *linear mode of operation*, and the other with $V_{ds} \geq V_{gs} - V_{T0}$ called the *saturation mode of operation* (and labeled so since the current in the MOSFET is saturated to an almost constant value). Thus the current voltage relation for the MOSFET can be written as

$$\begin{aligned} I_{ds} &= \mu C_{ox}\frac{W}{L}\left((V_{gs} - V_{T0})V_{ds} - \frac{V_{ds}^2}{2}\right), \quad V_{ds} < V_{gs} - V_{T0} \\ &= \frac{1}{2}\mu C_{ox}\frac{W}{L}(V_{gs} - V_{T0})^2(1 + \lambda V_{ds}), \quad V_{ds} \geq V_{gs} - V_{T0} \end{aligned} \tag{2.66}$$

Here λ is the channel length modulation parameter. The plot of I_{ds} versus V_{gs} is called the *transfer characteristics* of the MOSFET and the plot of I_{ds} versus V_{ds} is called the *output characteristics* of the MOSFET (Figure 2.12).

Sub-threshold Operation

In the text we have only discussed the current-voltage characteristics of the MOSFET operating above the threshold voltage, i.e. $V_{gs} > V_{T0}$. When $V_{fb} < V_{gs} < V_{T0}$ the MOSFET operates in the subthreshold mode. In this region, Poisson's equation can be written as

$$\frac{d^2\varphi}{dx^2} \approx \frac{qN_a}{\epsilon_S}\left(1 + e^{\frac{q\varphi}{kT}}\right) \tag{2.67}$$

The method of solving the equation is very similar to what we used for Equation 2.55. We find that in the case of subthreshold operation the free charge in the channel is given by

$$Q_f = \frac{C_{ox}kT\alpha}{q}e^{\frac{q(V_{gs}-V_{T0}-V_{ch}(y))}{1+\alpha}} \tag{2.68}$$

where $\alpha = \frac{1}{2C_{ox}}\left(\frac{q\epsilon_S N_a}{|\varphi_i|}\right)^{1/2}$. While determining the subthreshold current in the MOSFET it must be noted that the current due to diffusion is greater than the drift current due to the low carrier densities. Thus, the current is proportional to dQ_f/dy, and is given by

$$I_{ds} = \mu C_{ox}\alpha\left(\frac{kT}{q}\right)^2\frac{W}{L}e^{\frac{q(V_{gs}-V_{T0})}{1+\alpha}}\left(1 - e^{\frac{-qV_{ds}}{\alpha kT}}\right) \tag{2.69}$$

2.8 Conclusion

This chapter dealt with the basics of semiconductor devices. It introduced the terminology associated with semiconductors and introduced electrostatic analysis using Poisson's equation. Most importantly the chapter discusses MOSFETs. The concept behind MOSFETs is very similar to the thin film transistor in the future chapters. The unfamiliar reader must take some time and familiarise himself with these concepts which will prove to be the basis for the chapters ahead. The interested reader can consider the following texts for more information [3], and [4].

3

Circuit Analysis of MOSFET Circuits

CONTENTS

In this chapter we study the circuit aspects of the metal oxide semiconductor based field effect transistors (MOSFETs) which were introduced in the previous chapter. We study the circuit aspects of MOSFETs based on the MOSFET operation discussed in the previous chapter.

    We first introduce some simple MOSFET circuits, and then introduce a

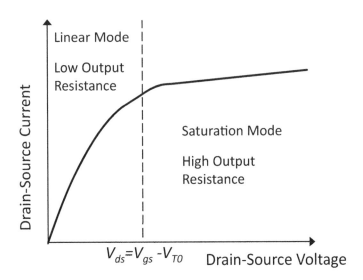

FIGURE 3.1
Output characteristics of the MOSFET.

small signal model for MOSFET circuit analysis. Using the small signal model we discuss its application in the analysis of single stage amplifiers. We then look at the parasitic capacitances that arise in a MOSFET and its importance in the frequency response and speed of MOSFET circuits. Finally, we conclude the chapter with an introduction to noise in electronic circuits.

3.1   MOSFET Operation and its Impact in Circuit Design

The output characteristics of the MOSFET is as shown in Figure 3.1.

For a gate-source voltage V_{gs} greater than the threshold voltage, V_{T0}, the MOSFET is in inversion mode of operation. As the drain-source voltage is varied from 0 to $V_{ds} > V_{gs} - V_{T0}$, the MOSFET changes its operation from the linear to saturation region.

In linear mode of operation $(0 < V_{ds} < V_{gs} - V_{T0})$, the current through the MOSFET is sensitive to the change in V_{ds} and the current voltage characteristics of the MOSFET in this region is given by

$$I_{ds} = 2\beta \left((V_{gs} - V_{T0})V_{ds} - \frac{V_{ds}^2}{2} \right)$$

(3.1)

Here $\beta = \frac{1}{2}\mu C_{ox}\frac{W}{L}$, where μ is the mobility, C_{ox} the dielectric capacitance per unit area, W the channel width and L the channel length. Therefore, in this region of operation, for small changes in V_{ds}, the MOSFET behaves like a resistor.

Once $V_{ds} > V_{gs} - V_{T0}$, the MOSFET moves into the saturation mode of operation. The current through the MOSFET is no longer strongly dependent on V_{ds} and instead mostly depends on variations in V_{gs}. The drain-source voltage does have a weak influence on the current through the MOSFET due to channel length modulation. The current-voltage characteristics in this region is given by

$$I_{ds} = \beta \left(V_{gs} - V_{T0}\right)^2 \left(1 + \lambda V_{ds}\right) \tag{3.2}$$

Here λ is the channel length modulation parameter. Since the current depends weakly on V_{ds}, the MOSFET almost behaves like a constant current source in this region of operation.

3.2   Small Signal Analysis of MOSFET Circuits

For the purpose of circuit analysis, it is useful to study how the MOSFET responds in different bias conditions for small fluctuations in the bias. This study is useful since it linearizes a non-linear device. In many practical applications such as amplifiers, an incoming signal from a sensor or any other source has to be amplified. The MOSFETs in these circuits are usually fixed at some bias voltages while the input signals are like small fluctuations riding on top of the "large signal" bias voltages. Here once again, the MOSFET can be expected to behave like a linear device and small signal analysis serves as a powerful tool for the analysis of these circuits. In this section we briefly discuss the methods to perform small signal analysis in MOSFET and MOSFET circuits.

When performing a small signal analysis, the dc large signal or dc nodes can be considered to be equivalent to ground nodes. This does not mean that the MOSFETs are biased at ground voltage. It simply implies that we are using the superposition principle - we are considering the small signal as an additional voltage or current source along with the large signal dc voltage or current sources. The role of the dc sources is to set the MOSFET at the desired bias point, while the small signal analysis tells us how the MOSFET behaves for signals around that bias point.

For the sake of clarity, we always assign small signal fluctuations in noncaps, eg., v and i are small signal voltages and currents while V and I are large signal or dc voltages and currents.

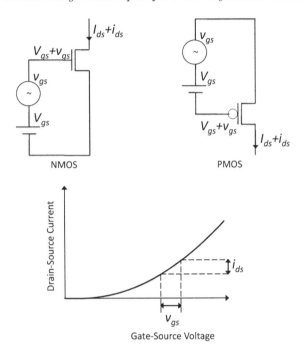

FIGURE 3.2
Small signal gate bias fluctuations.

3.2.1   Small Signal Gate Bias Fluctuations

First, let us consider a MOSFET with a large signal gate bias of V_{gs} and receiving a small signal fluctuation of $v_{gs}(t)$ on the gate. The effective gate voltage on the MOSFET is $V_{gs} + v_{gs}(t)$. The current through the MOSFET will also have a large signal component, I_{ds} corresponding to V_{gs} and a small signal fluctuation, $i_{ds}(t)$ because of $v_{gs}(t)$. This is illustrated in Figure 3.2. What is the influence of $v_{gs}(t)$ on $i_{ds}(t)$? We define the parameter $\frac{\partial I_{ds}}{\partial V_{gs}}$ called the *transconductance* of the MOSFET and denoted by g_m.

If the MOSFET is in linear mode of operation, V_{ds} is fixed and

$$g_{m,lin} = \frac{\partial I_{ds}}{\partial V_{gs}} = \frac{i_{ds}}{v_{gs}} = 2\beta V_{ds} \tag{3.3}$$

If the MOSFET is biased in saturation mode of operation with $V_{ds} \geq V_{gs} - V_{T0}$,

$$g_{m,sat} = \frac{\partial I_{ds}}{\partial V_{gs}} = \frac{i_{ds}}{v_{gs}} = 2\beta(V_{gs} - V_{T0})(1 + \lambda V_{ds}) \approx 2\beta(V_{gs} - V_{T0}) \tag{3.4}$$

For a MOSFET in saturation, $g_m \approx 2\beta(V_{gs} - V_{T0})$ and is equivalent to $g_m \approx 2(\beta I_{ds})^{1/2}$.

The relation between the small signal current fluctuation and gate bias can be written as

$$i_{ds} = g_m v_{gs} \tag{3.5}$$

Therefore, the transconductance parameter tells us about how "strong" the MOSFET is in converting a gate voltage change to an eqivalent drain-source current change.

3.2.2   Small Signal Drain Bias Fluctuations

Small voltage fluctuation in the drain-source bias also influence the drain-source current as shown in Figure 3.3. We can define a term $g_{ds} = \frac{\partial I_{ds}}{\partial V_{ds}}$. If the MOSFET is in linear mode of operation, V_{gs} is fixed and

$$g_{ds,lin} = \frac{\partial I_{ds}}{\partial V_{ds}} = \frac{i_{ds}}{v_{ds}} = 2\beta(V_{gs} - V_{T0} - V_{ds}) \tag{3.6}$$

If the MOSFET is biased in saturation mode of operation with $V_{ds} \geq V_{gs} - V_{T0}$,

$$g_{ds,sat} = \frac{\partial I_{ds}}{\partial V_{ds}} = \frac{i_{ds}}{v_{ds}} = \beta(V_{gs} - V_{T0})^2 \lambda \tag{3.7}$$

Since the resistance of the MOSFET is the change in current to the change in drain-source voltage, we find the *small signal resistance* or more precisely, the *output resistance*, or *output impedance*, since it is seen from the output terminals, of the MOSFET is $1/g_{ds}$. Thus, in linear mode of operation, the MOSFET resistance, r_{out} is given by

$$r_{out,lin} = \frac{1}{g_{ds,lin}} = \frac{1}{2\beta(V_{gs} - V_{T0} - V_{ds})} \tag{3.8}$$

As a matter of curiosity, in deep linear mode operation where $V_{ds} << V_{gs} - V_{T0}$, $r_{out,lin} \approx \frac{1}{2\beta(V_{gs}-V_{T0})} = \frac{1}{g_{m,sat}}$. For a MOSFET operating in saturation bias,

$$r_{out,sat} = \frac{1}{g_{ds,sat}} = \frac{1}{\beta(V_{gs} - V_{T0})^2 \lambda} = r_o \tag{3.9}$$

If the channel length modulation were negligible, $\lambda = 0$ and $r_{out,sat} = \infty$ implying that the MOSFET behaves like an idea current source. Note that we define the output impedance of a MOSFET in saturation as r_o.

3.2.3   Impedance Analysis of MOSFET Circuits

For any given circuit in a black box we can extract the effective impedance seen by a small time-varying signal at the input and output of the circuit.

In order to identify the input and output impedance, we apply an imaginary test voltage v_t and measure the current through the circuit i_t. This current is indicative of the effective impedance $Z = v_t/i_t$.

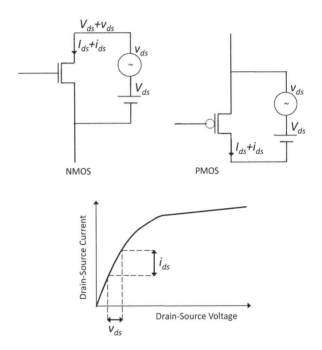

FIGURE 3.3
Small signal drain bias fluctuations.

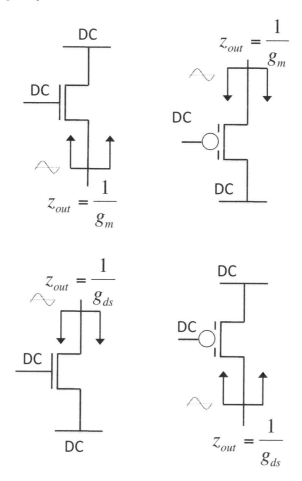

FIGURE 3.4
Small signal output impedance of the MOSFET.

In the case of a single MOSFET, the impedances as seen from the output terminals are shown in Figure 3.4. The values of these impedances vary depending on whether the MOSFETs are biased in the linear or saturation region, as described by Equation 3.3, Equation 3.4, Equation 3.8 and Equation 3.9.

3.2.3.1 Example

For example, let us consider the circuit shown in Figure 3.5 with two NMOS transistors and both biased in saturation. What is the output impedance of the circuit? We first consider all the dc nodes, i.e., power supply, and the gate

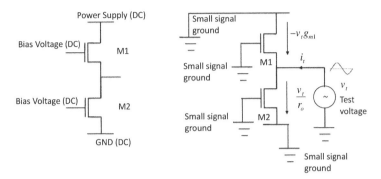

FIGURE 3.5
Example for output impedance calculations in a MOSFET circuit.

bias as ground. We then apply a test voltage, v_t at the output node of the circuit and study the current i_t.

Consider the influence of the test voltage on transistor M1. Since the source terminal of M1 is fluctuating and the gate-ground bias is fixed, we have an effective fluctuation in the gate-source voltage of M1. The small signal gate to source voltage fluctuation is $-v_t$. Therefore, this would cause a transistor current of $-g_{m1}v_t$ flowing into the test signal or equivalently a current of $g_{m1}v_t$ flowing out of the test signal source. Next, consider the influence of the test voltage on M2. Since the drain voltage is fluctuating, and since the MOSFET is biased in saturation, if we ignore channel length modulation, the fluctuations on the drain bias should not have any influence on the current through M2. However, including channel length modulation, we can consider the MOSFET M2 to behave like a resistor of resistance r_{o2} of Equation 3.9. The current through the MOSFET M2 due to the test voltage is therefore, v_t/r_{o2}. The total current, i_t, flowing into the circuit from the test signal source v_t at the output node is the sum of the currents through M1 and M2 and is given by

$$i_t = g_{m1}v_t + v_t/r_{o2} \tag{3.10}$$

The output impedance of the circuit is therefore v_t/i_t and is given by

$$Z_{out} = \frac{v_t}{i_t} = \frac{1}{g_{m1} + \frac{1}{r_{o2}}} = \frac{1}{g_{m1}}||r_{o2} \tag{3.11}$$

This implies that the output impedance is as though we have a parallel combination of resistances $1/g_{m1}$ and r_{o2}.

3.2.4   Transfer Function Analysis of MOSFET Circuits

Small signal analysis can be used to identify the influence of the input signal on the output signal. The ratio of the output signal to the input signal is called

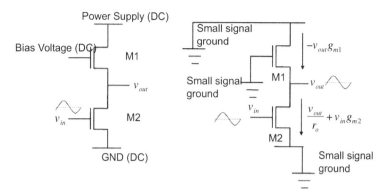

FIGURE 3.6
Example for gain or transfer function calculations in a MOSFET circuit.

the *transfer function* or *gain* of the circuit. For example in a single MOSFET, if the gate-source voltage is v_{in}, and the drain-source current, i_{out} is considered the output, we have the transfer function represented by $i_{out}/v_{in} = g_m$.

3.2.4.1 Example

Let us now consider the example of the preceding section, where the circuit had two NMOS MOSFETs biased in saturation shown again in Figure 3.6. Consider the input signal, v_{in}, appearing at the gate of M2 and the output voltage signal, v_{out} appearing at the drain of M2. What is the transfer function or gain, v_{out}/v_{in} of this circuit? The transfer function can be obtained by using *Kirchhoff's current law* at the output node of the circuit, i.e., summing all currents entering the output node to zero. The input v_{in} results in a current, $g_{m2}v_{in}$ through M2 flowing from the drain of M2 to ground and therefore out of the output node. The fluctuations in the output also results in currents through both M1 and M2. The current through M1 due to v_{out} is $-g_{m1}v_{out}$ flowing from the drain of M1 into the output node. The current through M2 due to v_{out} is v_{out}/r_{o2} which flows from the drain of M2 i.e. out of the output node. Equating the inflow and outflow of currents at the output node (the currents have nowhere else to flow)

$$g_{m2}v_{in} + \frac{v_{out}}{r_{o2}} = -g_{m1}v_{out} \qquad (3.12)$$

Therefore, the transfer function or small signal gain, A of this circuit is

$$A = \frac{v_{out}}{v_{in}} = -\frac{g_{m2}}{g_{m1} + \frac{1}{r_{o2}}} \qquad (3.13)$$

Since the output impedance calculated in the previous section is equal to $Z_{out} = \frac{1}{g_{m1} + \frac{1}{r_{o2}}}$, we have $A = -g_{m2}Z_{out}$.

3.3   High Frequency Response of MOSFET Circuits

As we saw in the first chapter, when a time-varying voltage is applied across a capacitor, we have a current through the capacitor. This current is proportional to the rate of change of the voltage across the capacitor, which is equivalent to the frequency of the small signal voltage across the capacitor.

In the previous section, when we estimated the impedance of MOSFET circuits, we never considered the capacitive components. This is perfectly fine if we assume that the small signals are having low frequency components thereby making the currents through the capacitors very small. However, as the frequency components in the small signals to the MOSFET circuits increase, the capacitive currents become dominant and it becomes essential to include these components in our estimates of the circuit impedances and transfer functions. In this section we first look at the capacitive components in a MOSFET and flow through some examples on the frequency analysis of MOSFET circuits.

3.3.1   Capacitive Components in a MOSFET

There are several capacitances associated with the standard MOSFET structure as shown in Figure 3.7. The doped contacts have an overlap length L_{ov} with the gate electrode, and an effective channel length of L in between them. The channel width of the MOSFET can be considered to be W.

The gate capacitance C_g, is a combination of the gate-source capacitance, C_{gs}, gate-drain capacitance, C_{gd} and the gate-bulk or body capacitance, C_{gb}. The gate-source capacitance is in turn the summation of the overlap capacitance between the gate and doped source electrode (C_{gs-ov}) and the gate-channel (C_{gcs}) capacitance near the source electrode, i.e., $C_{gs} = C_{gs-ov} + C_{gcs}$. The gate-drain capacitance is a summation of the overlap capacitance between the gate and drain electrode (C_{gd-ov}) and the gate-channel capacitance near the drain (C_{gcd}), i.e., $C_{gd} = C_{gd-ov} + C_{gcd}$. If the overlap length between the gate-source electrode and gate-drain electrode is the same, ignoring the effect of the fringing field lines, $C_{gs-ov} = C_{gd-ov} \approx C_{ox} W L_{ov}$.

When the MOSFET is in cut-off mode of operation, i.e. when there is no channel formation, $C_{gcs} = C_{gcd} = 0$, and $C_{gb} = (C_{ox}C_{dep}/C_{ox} + C_{dep})WL$, where C_{dep} is the depletion capacitance per unit area. Therefore, $C_{gs} = C_{gd} = C_{ox}WL_{ov}$.

In linear mode of operation, the channel is alomost equally distributed between the source and drain electrodes and shields the gate to body influence. Hence, $C_{gcs} = C_{gcd} = C_{ox}WL/2$, $C_{gb} = 0$ and therefore $C_{gs} = C_{gd} = C_{ox}WL/2 + C_{ox}WL_{ov}$.

In the saturation mode of operation, the channel is pinched off near the drain electrode and the charge near the source electrode is roughly two third the total charge in linear mode of operation. Thus, $C_{gcs} = 2C_{ox}WL/3$, $C_{gcd} =$

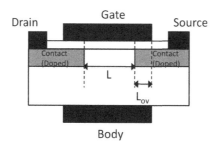

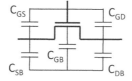

FIGURE 3.7
Capacitances associated with a MOSFET.

0, and $C_{gb} = 0$, and therefore, $C_{gs} = 2C_{ox}WL/3 + C_{ox}WL_{ov}$, and $C_{gd} = C_{ox}WL_{ov}$.

The interface between the strongly doped contacts and the bulk is a diode that has its intrinsic capacitance known as the *diffusion capacitance* and labeled as C_{sb} and C_{db}, for the source and drain, respectively.

It must be noted that all the channel capacitances are a function of the bias condictions, particularly gate bias.

3.3.2 Frequency Response

In order to estimate the frequency response of MOSFET circuits, one elegant approach is to break down the circuit into equivalent resistive and capacitive components. At each node of the circuit one can estimate the effective capacitance at the node, and the resistance of the signal path to the node. This associates each node of the circuit with the output node of some series RC circuit. The time constant associated with these nodes determine the frequency response of the circuit.

3.3.2.1 Miller Effect and Miller Capacitances

When establishing the capacitance associated with each node of a circuit, we may come across capacitances that are shunted across circuit blocks as shown in Figure 3.8. If the capacitance, C_s is shunted across a block with gain A, we must condider the effect of this gain block on the capacitance. This is called the *Miller Effect*.

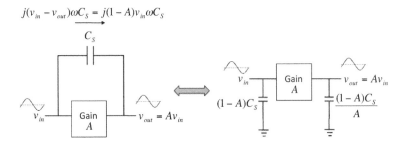

FIGURE 3.8
Miller effect and Miller capacitances.

Assume there is a small signal variation v_{in} at the input of this circuit, such that the output is $v_{out} = Av_{in}$. Looking from the input side, the capacitor has a voltage $v_{in} - Av_{in}$ across it causing a current of magnitude $i_{in} = v_{in}(1 - A)\omega C_s$ through the capacitor where ω is the frequency of the v_{in}. The source of the input signal now has to source a current of i_{in} through the capacitor. Therefore, the impedance caused by the capacitor as seen by the source is $Z_{in} = v_{in}/i_{in} = 1/(1 - A)\omega C_s$, which is equivalent to the impedance caused by a capacitor of capacitance $C_s(1 - A)$ between the input node and ground. On the output side, the current sourced (or sunk) by v_{out} is the same as i_{in}. Therefore, the output node of the circuit feels that it is seeing an effective impedance $Z_{out} = v_{out}/i_{in} = A/(1 - A)\omega C_s$, which is equivalent to the impedance of a capacitor of capacitance $C_s(1 - A)/A$ between the output node of the circuit and ground.

Therefore, looking from the input side, it appears as though C_s is equivalent to a capacitor of capacitance $C_s(1 - A)$ connected between the input node and ground. On the other hand, looking from the output side, it appears that there is a capacitor $C_s(1 - A)/A$ connected between the output node and ground. Therefore, we can break C_s into two equivalent capacitances — one at the input node and ground and having a magnitude $C_s(1 - A)$ and the other between the output node and ground having a magnitude $C_s(1 - A)/A$. The effect of the gain block on the effective capacitance is called the *Miller effect* and these equivalent capacitances are called the *Miller capacitances*.

Though we have considered only capacitive components in this discussion, this is valid for any general impedance component shunting a gain block.

3.3.2.2 Example

Consider the example of the circuit shown in Figure 3.9. The source of the small signal v_{in} has an output impedance of R_{in}. First, ignoring the capacitive components of the circuit, i.e., performing a low frequency small signal

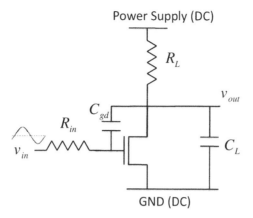

FIGURE 3.9
Example for high frequency gain calculations in a MOSFET circuit.

analysis, the output impedance of the circuit is seen to be

$$R_{out} = \frac{1}{\frac{1}{R_L} + \frac{1}{r_o}} = R_L || r_o \tag{3.14}$$

The gain or transfer function of the circuit is seen to be

$$A_{lf} = \frac{v_{out}}{v_{in}} = \frac{g_m}{\frac{1}{R_L} + \frac{1}{r_o}} = -g_m R_{out} \tag{3.15}$$

If we now consider high frequency inputs to the circuit, the capacitive components can no longer be ignored. We now try to associate capacitances to the different nodes of the circuit. There are two primary nodes — the input and the output node. Noting that C_{gd} is a shunt between the input node and the output node, we can find the equivalent Miller capacitances associated with the input and output nodes. Since the low frequency gain of the circuit is $-g_m R_{out}$, we find the input Miller capacitance associated with C_{gd} is $(1 + g_m R_{out})C_{gd}$ and the equivalent Miller capacitance associated with the output node is $(1 + g_m R_{out})C_{gd}/g_m R_{out}$. Thus, the total capacitance seen at the input is

$$C_{in} = (1 + g_m R_{out})C_{gd} + C_{gs} \tag{3.16}$$

The total capacitance seen at the output is

$$C_{out} = \frac{(1 + g_m R_{out})C_{gd}}{g_m R_{out}} + C_L \tag{3.17}$$

The total effective resistance in the signal path towards the input is R_{in} while the total effective resistance in the signal path towards the output node is R_{out}. Thus, the effective time constants associated with the input and output node of the circuit are $R_{in}C_{in}$ and $R_{out}C_{out}$, respectively. Therefore, the transfer function at high frequency, also known as the high frequency gain of the circuit is given by

$$A_{hf} = \frac{A_{lf}}{(1 + j\omega R_{in}C_{in})(1 + j\omega R_{out}C_{out})} \tag{3.18}$$

3.4 Noise in Circuits

Electronic circuits based on semiconductors and passive components like resistors involve the motion of electrons. The electron movement within the material results in the interaction of the electrons with the lattice and defects in the material resulting in random fluctuation of the macroscopic quantities such as the voltage across the circuit element or current through the circuit element. These random fluctuations with respect to electronic circuits constitute electronic noise.

While the general behavior of the circuits we deal with in this book are not greatly influenced by noise, the analysis of noise itself is an important aspect of circuit design — particularly when it comes to sensing small signals. From the point of view of large-area electronics, which is the scope of this book, noise is an important component for the design of large area electronic sensors such as pin diode based image sensors.

In this section we briefly look at the representation of noise in circuits, the types of noise and their physical origins, and finally noise analysis within some circuits.

3.4.1 Representation of Noise

Random fluctuations in current and voltage are what consitute electrical noise. How do we characterize this noise?

3.4.1.1 Probability Density Function

Since electronic noise is a random process, its frequency, phase, and amplitude are a part of some probability density function. Depending on the nature of this function, and the physical processes that create them, there are different types of noise. If the noise — which is a random process — has a density function that does not change with time, it is said to be a *stationary process*. The values of the mean, variance, and other moments of the process will also be time invariant.

3.4.1.2 Noise Power

An obvious characteristic of noise is the average amplitude or power contained in the noise component. For a voltage fluctuation $v(t)$ across a resistor R, the power generated or dissipated by the noise alone is the time average of $v(t)^2/R$. Since $v(t)$ is a random signal (which we assume to be stationary), we estimate its average by its mean square average value, which is

$$\langle v^2 \rangle = \lim_{T \to \infty} \frac{1}{T} \int_0^T v(t)^2 dt \qquad (3.19)$$

Since the mean power dissipated is $\langle v^2 \rangle/R$, the term $\langle v^2 \rangle$ is also called, perhaps misleadingly, as the *noise power*. Thus, we can have a voltage noise power $\langle v^2 \rangle$ with units of volt^2 and a current noise power of $\langle i^2 \rangle$ measured in ampere^2. The square root of the noise power provides the equivalent root mean square average value of the noise in volts and amperes.

3.4.1.3 Power Spectrum

Noise not only has a random distribution of amplitudes, but a random distribution of frequency components as well. In order to characterize the aspect of frequency content in electronic noise we consider the *power spectrum* of noise. The power spectrum of noise is the plot of the average power of the noise at every frequency, i.e., a plot of the noise power versus frequency. The power spectrum of voltage and current noise has units of volt^2/Hz and amperes^2/Hz, respectively. Equivalently, the measure of the noise component in root mean square terms at every unit frequency band is volt/Hz$^{1/2}$ and ampere/Hz$^{1/2}$, respectively.

In the first chapter we discussed the filtering properties of RC circuits. When a signal of many frequencies was sent to an RC circuit, all the frequencies above the cut-off frequency were removed or diminished, while the frequency components of the signal below the cutoff frequency survived intact. So what happens if say a voltage noise is sent through an RC circuit? Since voltage noise, or any other noise is really and physically an electronic signal (though random), it too will be filtered by the RC circuit. In fact, if we send noise through an arbitrary black box, which may be composed of any electrical circuit, the noise will be influenced by the filtering properties of that black box. If the general transfer function of the black box, which is the output to input signal ratio is $T(f)$, and the spectrum of the noise submitted to the black box is $S_{in}(f)$, the spectrum of the noise at the output of the black box is $S_{out}(f) = S_{in}(f)|T(f)|^2$.

We will revisit this important concept once we discuss the different types of noise.

3.4.2   Types of Noise

3.4.2.1   Thermal Noise

Thermal noise is the noise generated due to the Brownian motion associated with the flow of electrons in a conductor. It is also called the *Johnson noise*. An important concept while considering thermal noise is that there be a "continuous" flow of electrons through the circuit element. Thus, the element must be conductive or resistor like.

When a voltage V is applied across a resistor with resistance R, we have a current that is a sum of the large average current V/R as expected by Ohm's Law and the current due to the thermal noise current. On the other hand, when a constant current, I is sent through the resistor, it generates a voltage across it which is the sum of IR and the thermal noise voltage. The noise is therefore represented as a voltage noise source in series with the noiseless resistor, or a current noise source in parallel with a noiseless resistor.

The thermal noise current through a resistor of resistance R has a power spectrum defined by

$$S_i = 4kT/R \tag{3.20}$$

Here k is the Boltzmann's constant, T is the temperature. This is the current noise power per unit frequency band and has the units of ampere^2/Hz. The thermal noise voltage through a resistor is

$$S_v = S_i R^2 = 4kTR \tag{3.21}$$

and has units volts^2/Hz.

Note that the noise is independent of the voltage applied or the current through the resistor. Moreover, the noise spectrum is independent of the frequency, which implies that the thermal noise has the same noise power at all frequencies. Therefore, it is also called *white noise*. In reality however, there cannot exist an ideal white noise source. Nevertheless, if the noise component has a constant power spectrum in the frequency band of interest it is termed "white".

We now consider the question asked in the previous section. What happens when noise is filtered through an RC circuit. Since the RC circuit has a resistor in series with the capacitor, the resistor generates a thermal noise voltage, which is filtered through the circuit. The square amplitude of the transfer characteristics of the RC circuit is

$$|T(f)|^2 = \left| \frac{V_{out}}{V_{in}}(f) \right|^2 = \left| \frac{1}{1 + j2\pi fRC} \right|^2 = \frac{1}{1 + R^2C^2f^2} \tag{3.22}$$

The thermal noise due to the resistor acts as the input noise to the RC circuit with $S_{in} = 4kTR$ per unit frequency. The output noise, S_{out}, which is the filtered version of S_{in}, is measured across the capacitor. The per unit frequency output noise is

$$S_{out}(f) = S_{in}(f)|T(f)|^2 = 4kTR\frac{1}{1 + 4\pi^2 R^2C^2f^2} \tag{3.23}$$

The total noise power at the output is the summation of $S_{out}(f)$ over all frequencies

$$
\begin{aligned}
P_{out} &= \int_0^\infty 4kTR\frac{1}{1+4\pi^2R^2C^2f^2}df \\
&= kT/C
\end{aligned}
$$
(3.24)

This noise component due to an RC circuit is therefore called kT/C *noise*. Notably, it is independent of the resistance of the resistor being used.

3.4.2.2 Shot Noise

In electronic devices where the electronic conduction is by discrete packets of charge, we observe noise due to number fluctuations in these charge packets per unit time. It is this variance from one unit time to the next that is the source of noise. Thus, when the electrons or electron packets through a circuit are few and spread out in time and are transmitted independent of each other, we observe *shot noise* e.g., current flow due to photoelectric excitation in p-n diodes. For shot noise to exist there must be current flow.

The power spectrum of shot noise is proportional to the average current, I_{avg}, through the device and is given by

$$
S_i = 2qI_{avg}
$$
(3.25)

where q is the electron charge. Shot noise is characterized by a white noise spectrum up to a certain cut off frequency, which is related to the transit time of the electron through the conductor. In contrast to thermal noise, shot noise is not temperature dependent.

Shot noise is not easily measured in good conductors because the current fluctuations resulting due to the discreteness of electrons are smoothed out. However, as we get to nanoscale devices that have few carriers through them, shot noise can be measured.

3.4.2.3 Flicker Noise

Flicker noise is one of the most ubiquitous noise sources in the universe. Many phenomena ranging across market prices, earthquakes, currents in conductors, etc, have a flicker noise component. The primary feature of flicker noise is that the power spectrum of the noise has a $1/f$ dependence on the frequency. Thus, the noise power is extremely large at low frequencies and very small at high frequencies. Due to the larger presence in the lower frequency of the spectrum the noise has a "red" or "pink" component to it and is called "pink" noise.

The voltage noise power spectrum of flicker noise is given by

$$
S_v \propto 1/f
$$
(3.26)

Generally, noise components with a $1/f^\lambda$ dependence on frequency, with λ being some constant coefficient, are also considered as flicker noise.

In electronic devices, flicker noise is attributed to carrier number fluctutations and carrier mobility fluctuations.

Since flicker noise has a $1/f$ dependence, and since thermal noise is invariant with frequency, there occurs a certain frequency, f_c beyond which thermal noise dominates and before which flicker noise dominates. This frequency is called the *corner frequency* and is defined as the frequency at which the noise power density of flicker noise equals the power density of thermal noise. Identifying the corner frequency is useful to identify the noise we are dealing with given the operational frequency of our circuit.

3.4.3   Noise in Field Effect Transistors

3.4.3.1   Thermal Noise

Thermal noise is present in MOSFETs as well provided the transistor is on and there is a conductive channel (due to inversion) between drain and source. Since the small signal resistance of a MOSFET is $1/g_m$, the thermal noise current in a MOSFET is

$$S_i = 4kT\alpha g_m \qquad (3.27)$$

where α is some device dependent parameter. For long channel devices $\alpha \approx 2/3$, and we assume this value of α throughout.

When the MOSFET is used as a switch along with a capacitor, it forms an RC circuit. As discussed earlier, the RC circuit produces a kT/C noise across the capacitor and this noise is independent of the resistance of the resistor being used. The filtered noise of the MOSFET present on the capacitor has a total power of kT/C volt^2.

3.4.3.2   Shot Noise

Shot noise is not considerable or easily measured in the transistor current during MOSFET operation due to the continuous mode of electron current transport (i.e., not in discrete independent packets). Therefore, we ignore this component of noise during noise analysis with MOSFET circuits.

However, if the circuit contains components such as Schottky diodes, photodiodes, phototransistors, etc, where the physical mechanism is due to discrete packet like transport, shot noise, becomes an important component of the noise analysis.

3.4.3.3   Flicker Noise

Flicker noise in MOSFETs is attributed to carrier number fluctations and carrier mobility fluctuations, primarily due to the presence of dangling bonds at the semiconductor insulator interface.

The voltage noise power spectrum of flicker noise in MOSFETS is given by

$$S_v = \frac{\gamma}{C_{ox}WLf} \qquad (3.28)$$

The constant of proportionality γ is process dependent and is typically $1e - 26V^2F$. Notably, the flicker noise in MOSFETs can be reduced by scaling up the area by increasing the channel width and channel length — particularly for the input stage of the circuit. The equivalent flicker noise current in MOSFETs is given by

$$S_i = \frac{\gamma}{C_{ox}WLf}g_m^2 \tag{3.29}$$

The corner frequency for MOSFETs is defined by equating the flicker noise density to the thermal noise density,

$$\frac{\gamma}{C_{ox}WLf_c}g_m^2 = 4kT\alpha g_m \tag{3.30}$$

This results in $f_c = \frac{\gamma}{4kTC_{ox}WL\alpha}$.

3.4.4 Noise in MOSFET Circuits

3.4.4.1 Representing Noise in MOSFET Circuits

Any general MOSFET circuit has noise due to the noise components in the individual MOSFETs. Consider a MOSFET circuit represented as a black box. The circuit has an input impedance Z_{in}, gain A, an output resistance R_{out} and drives a capacitor load C_L. Let the summation of all the MOSFET noise components at the output is $\langle v_{no}^2 \rangle$.

The noise component at the output can be represented as an equivalent input noise voltage and current component, $\langle v_{ni}^2 \rangle$ and $\langle i_{ni}^2 \rangle$, respectively. We have the following relations,

$$\langle i_{no}^2 \rangle = \frac{\langle v_{no}^2 \rangle}{Z_{out}^2} \tag{3.31}$$

$$\langle v_{ni}^2 \rangle = \frac{\langle v_{no}^2 \rangle}{A^2} \tag{3.32}$$

$$\langle i_{ni}^2 \rangle = \frac{\langle v_{ni}^2 \rangle}{Z_{in}^2} = \frac{\langle v_{no}^2 \rangle}{(AZ_{in})^2} \tag{3.33}$$

The noise representation at the input of the circuit must include both the voltage and current noise components. This is to ensure that the noise input to the circuit under consideration is valid for any output impedance of the preceeding stage, which is the source of noise.

The output noise of the circuit is filtered by the RC circuit with resistance R_{out} and capacitance C_L. The total output noise power observed across C_L is given by

$$\langle v_{no,total}^2 \rangle = \int_0^\infty \frac{\langle v_{no}^2 \rangle}{1 + 4\pi^2 R_{out}C_L^2 f^2} df \tag{3.34}$$

For any input signal V_{in} to the noiseless circuit, the output signal has a magnitude AV_{in}. For the circuit under consideration, the presence of noise reduces

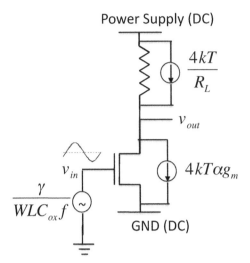

FIGURE 3.10
Noise analysis in a common source amplifier.

the signal–to–noise ratio (SNR). The SNR is thus given by the ratio of the signal power (which is the square of the amplitudes) at the output to the noise power at the output,

$$SNR = \frac{A^2 v_{in}^2}{\langle v_{no,total}^2 \rangle} \tag{3.35}$$

3.4.4.2   Example

We now consider the calculation of noise in some typical circuits. We consider the common source amplifier shown in Fig 3.10.

The common source amplifier is used as a primary gain element in circuits. The gain of the circuit is $A = -g_m(R_L||r_o) \approx -g_m R_L$ and the output impedance of the circuit is $Z_{out} = R_L||r_o \approx R_L$. The sum of all the current noise components at the output is

$$\langle i_{no}^2 \rangle = 4kT\alpha g_m + \frac{\gamma}{C_{ox}WLf} \frac{A^2}{Z_{out}^2} + \frac{4kT}{R_L} \tag{3.36}$$

Note that the flicker noise component $\frac{\gamma}{C_{ox}WLf}$ is the input referred voltage noise to the MOSFET. Hence, the output referred voltage noise to the MOSFET is the input noise times the square of the gain, and the output referred current is the output voltage noise by the square of the output impedance.

The total input referred voltage noise is $\langle v_{ni}^2 \rangle = \langle i_{no}^2 \rangle Z_{out}^2/A^2$ and is therefore given by

$$\langle v_{ni}^2 \rangle = \frac{4kT\alpha}{g_m} + \frac{\gamma}{C_{ox}WLf} + \frac{4kT}{g_m^2 R_L} \tag{3.37}$$

3.5   Conclusion

This chapter covered the basic ideas involved in the analysis of circuits based on MOSFETs. We have ignored the influence of the body contact in the MOSFET analysis with the purpose that these techniques readily transfer to the analysis of thin film transistor based circuits in the chapters ahead. The reader can consider the following texts for more information [5], [6], and [7].

Part II

Non Crystalline Semiconductors

4

Non-Crystalline Semiconductors

CONTENTS

In this chapter we briefly introduce non-crystalline semiconductors — what they are, why we need them, the materials used, and the electronic properties of these semiconductors. We touch these concepts very briefly in this chapter, and head towards the problems that influence circuit design with non-crystalline semiconductor based field effect transistors in the chapters ahead.

4.1   Introduction to Non-Crystalline Semiconductors

In order to define non-crystalline semiconductors, we first look at a crystalline semiconductor and consider the example of silicon. Crystalline silicon (mono–crystalline silicon) is generally grown slowly from a melt of silicon with a template crystal seed. Since the silicon atoms have enough energy and time, they move to the least energy positions resulting in a continuous lattice of diamond cubic cells with long range order. On the other hand, when the semiconductor material is rapidly cooled from a melt, or is deposited at low temperatures, the atoms or molecules of the semiconductor do not have sufficient time or energy to settle in the least energy position. The resulting *amorphous* or *glassy* film therefore does not have long range order, and contains numerous defects such as broken bonds, stretched bonds, and variations in bond angles. The result is a lack of long range order, with any possible lattice ordering existing for short lengths comparable to a few atoms long. These semiconductors are classified as *non-crystalline semiconductor*. Between these extremes of completely ordered crystalline semiconductors and completely disordered non-crystalline semiconductors we can have poly-crystalline semiconductors. For example, poly-crystalline silicon consists of a many single crystal "flakes" or "grains" called *crystallites* set together in a matrix with amorphous silicon in-between these grain boundaries. The crystallites need not have the same crystal geometry and are generally of the order of a few micro-meters in size. Thus, poly-crystalline silicon has much shorter order length. Our primary interest in this book is in glassy or amorphous semiconductors [9]-[12].

Materials used for non-crystalline semiconductors can be broadly classified as inorganic and organic. Inorganic semiconductors include amorphous silicon, amorphous metal oxides such as zinc oxide, galium-indium-zinc oxide, etc. One of the most well-studied inorganic non-crystalline semiconductor has been hydrogenated amorphous silicon (a-Si:H). This involved the understanding the gap states in hydrogentaed amorphous silicon [12]-[19], the understanding of conductance [25]-[21], and the modeling of hydrogenated amorphous silicon based field effect transistors [22]-[37]. Organic semiconductors can be single molecules (e.g., anthracene, rubrene), short chains called *oligomers* or *long chain polymers* (e.g., poly(3,3-dialkyl-quaterthiophene (PQT), poly(3-hexylthiophene) (P3HT)). The use of organic materials as a semiconductor for the synthesis of field effect transistors was considered several decades ago [39]-[46]. Since then there have been several areas of research [38], [47] involving studies on the electronic transport [48]-[53], fabrication techniques [54]-[66], modeling [67]-[83], materials [39], [43], [65], [84]-[93], and applications [94]-[98]. On the materials front it is noteworthy that both n-type (electron transporting) and p-type (hole transporting) polymers have been synthesized.

Non-crystalline semiconductors are fabricated on a variety of substrates such as glass, plastics, etc. Typical methods of fabrication of non-crystalline semiconductors includes using techniques such as chemical vapor deposition and sputtering for inorganic semiconductors, and solution processed techniques such as spin-coating and ink-jet printing for organic polymer semiconductors.

The main application of non-crystalline semiconductors is in fabricating electronics over large spatial areas, e.g., displays, x-ray image sensors. The use of crystalline silicon wafers are an extremely expensive proposition with problems such as mismatch. In such applications, non-crystalline semiconductors are more favorable due to their promise of lower fabrication costs, fabrication at lower temperature over large areas, and larger spatial uniformity. The possibility of low temperature fabrication of electronics also promises electronics on flexible substrates such as plastics [99]-[103].

4.2 Structure and Electronic Transport

In this section we study structural and electronic properties of inorganic and organic polymer non-crystalline semiconductors.

4.2.1 Inorganic Semiconductors

We use the concepts and ideas developed in understanding amorphous hydrogenated silicon to explain some general properties of inorganic disordered semiconductors.

Due to the non-crystalline nature of the semiconductor resulting in varying bond lengths, bond angles and dangling bonds, the band gap of the semiconductor is not free of states. All the unique properties in non-crystalline semiconductors, which make them differ in behavior from crystalline semiconductors is primarily due to the presence of states in the gap. In the case of crystalline Si, we have a clear band gap separating the conduction and valence bands. However, in the case of a-Si:H, the presence of states in the band gap results in a cruder definition of the conduction and valence band as shown in Figure 4.1.

The states in the band gap of a-Si:H that lie above the intrinsic Fermi level are called *acceptor-like states* that are charged when filled and neutral when empty. On the other hand, the states below the Fermi level are the *donor-like states* that are neutral when filled and charged when empty. In the case of a-Si:H there are more donor-like states than acceptor-like states and therefore the equilibrium Fermi level lies above midgap. This has significant consequences from the point of view of circuit design. Since it takes less charge to pull the Fermi level closer to the conduction band edge as compared to the valence band edge, an n-channel transistor is more feasible (has a higher on current) at low bias voltages as compared to a p-channel transistor.

The states in the gap can be further classified as *deep states* and *band-tail states*. The states that lie around the intrinsic Fermi level at the middle of the gap are called the deep states while the states nearer the conduction and valence band edges are called the band-tail states. The deep states are primarily due to the presence of broken or dangling bonds. Thus, the hydrogenation of amorphous silicon reduces the deep state density. The band-tail states are due to the presence of weak bonds in the a-Si:H structure. Density of states measurements have shown that the band-tail density of states in a-Si:H have an exponential distribution.

Carrier transport mechanisms in a-Si:H can be broadly divided into transport in the deep states, band-tail states, and in the extended states. Just as in crystalline semiconductors, thermal excitation of carriers from the Fermi level to above the conduction band edge boosts extended state transport. However, due to the presence of traps, the effective conductivity of carriers is low since

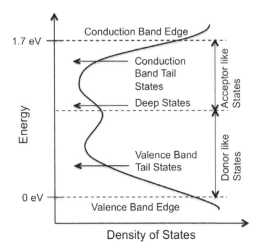

FIGURE 4.1
States in the band gap of non-crystalline semiconductors.

the carriers spend some time in the trap states. This *trap limited transport* mechanism is common to all non-crystalline semiconductors. *Hopping* is the main mechanism of transport for carriers in band-tail and deep states. However, hopping in band-tail states is strongly dependent on temperature, while hopping of carriers in deep states have a weaker temperature dependence. If the deep state density is significant, deep state hopping could become a significant mode of carrier transport at low temperatures. However, the passivation of dangling bonds with hydrogen results in negligible deep state conduction in a-Si:H.

4.2.2   Polymer Semiconductors

In order to understand electronic conduction in polymers we take the examples of two polymers — one having sp^3 hybridization and the other having sp^2 hybridization. In the case of the sp^3 hybridized polymer, the valence electrons of the molecules are bound in sp^3 hybridized covalent bonds therefore making the polymer non-conductive. However, in the case of the sp^2 hybridized polymer chain, the valence electrons of each center reside in the p_z orbital (pi orbital), which "floats" in a direction perpendicular to the polymer chain. These orbitals are *delocalized* and these polymers can be doped by adding (n-type) or removing (p-type) delocalized electrons by reduction or oxidation. Doping of these polymers create free charge carriers thereby allowing charge transport along the chain. When these polymers have strong *conjugation* or pi orbital overlaps, the band gap of the material is not very large, and the material can be made semiconducting.

The structure of polymer semiconductors can be described as poly-crystalline cells separated by amorphous regions. Polymers such as PQT can be modeled as a stack of sheets or lamelae with each lamelae having arrays of polymer chains.

Charge transport in polymer semiconductors is due to a combination of intramolecular transport, i.e., transport of carriers in a single molecule along the pi orbital overlaps and intermolecular hopping from one polymer chain to another. Charge transport along the chain is limited by the interaction of the carriers with trap states. The carriers spend some time in the trap states and the effective mobility is dependent on this time length. The charge transport between the chains is generally the limiting mechanism and is via hopping between chains. In the presence of an electric field ξ, the effective hopping conduction between the polymer chains increases in proportion to $exp(\alpha\xi^{1/2})$ in accordance with the *Poole Frenkel Mechanism*. Here α is a constant. An important feature in soft materials like polymers is the formation of *polarons*. When a electron moves through a lattice, it interacts with the neigboring atoms by pushing the electron cloud of these atoms away from it and thereby skewing the potential profile. In the case of polymers, the chain is more deformable and the charge gets trapped in a slightly deeper potential well. Therefore, charge transport in organic polymers are better described by considering the mobility of the polaron, which effectively implies the correct calculations of the effective mass of the carrier.

4.3 Thin Film Transistors

Electronic systems such as displays and image sensors required large area deposition of non-crystalline semiconductor films. The fundamental building block of these systems has been the field effect transistor based on non-crystalline semiconductors, also called *thin film transistors* (TFTs).

While there exist several geometries and process flows, a typical structure of the TFT consists of a stack of insulator and semiconductor deposited over a planar gate electrode as shown in Figure 4.2. The source and drain electrodes are then deposited over the semiconductor, with doping (if feasible) at the contacts and some overlap with the gate electrode to ensure low contact resistance. For a-Si:H based TFTs, a typical process flow is as follows. A gate metal layer of about 100 nm is first patterned on a substrate (usually glass or plastic). A stack of about 200 nm dielectric (typically silicon nitride), and about 100 nm of a-Si:H is then deposited with n+ doped a-Si:H at the source-drain contacts. While Si oxides form a good dielectric for crystalline Si based field effect transistors, Si nitride is the dielectric of choice in the case of a-Si:H due to the relatively defect free interface properties. The gate dielectric

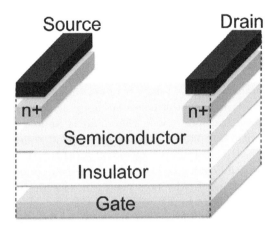

FIGURE 4.2
Structure of the thin film transistor.

vias are then patterned and etched after which source and drain contacts are patterned.

For solution processable organic polymer semiconductors the TFTs are typically fabricated as follows. After the deposition of a gate electrode by physical vapor deposition or ink-jet printing, a polymer insulator layer e.g., Polymethyl methacrylate (PMMA), poly-4-vinylphenol (PVP), is spin-coated. The substrate is then baked allowing the polymers in the insulator layer to cross-link. The source and drain electrodes are then deposited by physical vapor deposition or ink-jet printing. Finally the channel region is treated to make it compatible for "good" polymer semiconductor deposition. For example, in the case where PQT semiconductor is used, a self assembled monolayer of Octadecyltrichlorosilane (OTS) is grown in the channel region prior to the deposition of PQT. After PQT deposition (by ink-jet printing or spin-coating), the substrate is baked to the prescribed temperature (about 120C for 1 hour in the case of PQT). Since polymer semiconductors are generally very sensitive to the environment, particularly oxygen, a passivation layer is deposited over the semiconductor.

4.4 Conclusion

This chapter briefly introduced the concept of non–crystalline semiconductors. In the next chapter we look at the device physics of TFTs based on non-crystalline semiconductors.

5

Device Physics of Thin Film Transistors

CONTENTS

In the previous chapter, we introduced the semiconductor materials and typical fabrication processes of non-crystalline semiconductor based thin film transistors (TFTs). From this point on, whenever we mention TFT, we automatically imply that it is based on non-crystalline semiconductors.

In this chapter we focus on the device physics of TFTs. We establish the basic ideas of device physics, and the means to treat the non-crystallinity of the semiconductor using hydrogenated amorphous silicon (a-Si:H) based TFTs as an example.

5.1 Density of States in Non-Crystalline Semiconductors

The density of states in the band gap can be studied using techniques like field effect measurements [104], transient and steady-state photoconductivity [105], deep level transient spectroscopy [106], and capacitance-voltage measurements.

As discussed in the previous chapter, at thermal equilibrium, the states above the Fermi level are called acceptor-like since they are charged negative when filled with electrons and neutral when empty. Equivalently, the states below the Fermi level are donor-like since they are neutral when filled with electrons and charged positive when empty. Due to charge neutrality of the semiconductor, the Fermi level at thermal equilibrium, which is called the intrinsic Fermi level, E_i, sits at a location "exposing" just the right number of acceptor states and "covering" (thereby filling with electrons) a certain number of donor states to provide charge balance and hence neutrality.

5.1.1 Exponential Density of States

The states close to the middle of the energy gap are called deep states, while the states closer to the band edges are called band-tail states. In the case of certain non crystalline semiconductors (eg. a-Si:H), the density of states have an exponential profile such that

$$g_A(E) = g_{tA}e^{\frac{E-E_c}{kT_{tA}}} + g_{dA}e^{\frac{E-E_c}{kT_{dA}}} \tag{5.1}$$

$$g_D(E) = g_{tD}e^{\frac{E_v-E}{kT_{tD}}} + g_{dD}e^{\frac{E_v-E}{kT_{dD}}}$$

Here, the g_A and g_D are the density of acceptor-like and donor-like states, respectively. g_{tA} and g_{tD} represent the density of acceptor-like tail states close to the conduction band edge at E_c, and the donor-like tail states close to the valence band edge E_v, respectively. g_{dA} and g_{dD} represent the density of acceptor-like deep states and donor-like deep states. T_{tA}, T_{tD}, T_{dA}, and T_{dD} represent the characteristic temperatures (and equivalently energy) of the acceptor-like and donor-like tail states, and acceptor-like and donor-like deep states, respectively.

5.1.2 Trapped Charge Density

If the Fermi level moves above (towards E_c) its intrinsic level position, the total trapped charge (electrons) density in the acceptor-like states is given by

$$n_t = \int_{E_c}^{E_i} g_A(E)f(E)dE \tag{5.2}$$

If the Fermi level moves below (towards E_v) its intrinsic level position, the total trapped charge (holes) density in the donor-like states is given by

$$p_t = \int_{E_i}^{E_v} g_D(E)(1 - f(E))dE \tag{5.3}$$

Substituting for the Fermi function $f(E) \approx e^{-(E-E_f)/kT}$, in the above expressions, and defining $\varphi = E_f - E_i$, we obtain

$$
\begin{aligned}
n_t &= n_{ti}e^{\frac{\varphi}{V_{tA}}} + n_{di}e^{\frac{\varphi}{V_{dA}}} \\
p_t &= p_{ti}e^{-\frac{\varphi}{V_{tD}}} + p_{di}e^{-\frac{\varphi}{V_{dD}}}
\end{aligned}
\tag{5.4}
$$

Here, V_{tA}, V_{tD}, V_{dA}, and V_{dD} are the characteristic voltages related to the acceptor-like and donor-like tail states, and acceptor-like and donor-like deep states, respectively.

5.1.3 Free Charge Density

The free charge density in the semiconductor can be found by identifying the density of states in the conduction band (above E_c), say $g_c(E)$ and the density of states in the valence band (above E_v), say $g_v(E)$. The density of free electrons and holes are then given by

$$
\begin{aligned}
n_f &= \int_{E_c}^{\infty} g_c f(E) = n_{fi}e^{\frac{\varphi}{V_{th}}} \\
p_f &= \int_{-\infty}^{E_v} g_v f(E) = p_{fi}e^{\frac{-\varphi}{V_{th}}}
\end{aligned}
\tag{5.5}
$$

where $V_{th} = kT/q$ is the thermal voltage.

5.2 Device Physics of TFTs

Basically TFTs are field effect devices with a working principle not far from the crystalline semiconductor based MOSFETs discussed earlier. The primary differences in operation are due to the presence of states in the gap of the semiconductor. These states determine the charge transport in the semiconductor and the movement of the Fermi level with applied gate bias.

5.2.1 MOS Capacitor

Let us first consider the MOS capacitor based on non-crystalline semiconductors. The structure is exactly the same as a crystalline silicon (c-Si) based MOS capacitor, with the exception that the semiconductor now has states in the band gap. Initially, let us assume that the gate voltage is at the flat-band voltage and there is no band bending.

5.2.2   Forward Subthreshold Operation

As the gate voltage is increased a little, the Fermi level near the semiconductor-insulator interface begins to move towards the conduction band edge (see Figure 5.1). However, the presence of the large number of states in the middle of the band gap, pin the Fermi level and all the charge in the gate electrode is compensated by the charge trapped in the deep states. There is very little induced free charge in the semiconductor. This region of operation is the *subthreshold region* of operation. Moreover this region of operation is equivalent to the depletion mode of operation of the c-Si MOSFET. However, note the subtle difference here. Unlike the case of the c-Si MOS capacitor, there is no "artificial" dopants to reflect the charge on the gate electrode. The band bending in the semiconductor results in trapped charges in the already existing trap states in the semiconductor.

If the potential profile in the x-direction is $\varphi(x) = E_f - E_i(x)$, the variation of density of charge trapped in the deep states is $n_{di}e^{\frac{\varphi}{V_{dA}}}$. Poisson's equation under this situation can be written as

$$\frac{d^2\varphi}{dx^2} = \frac{qn_{di}e^{\frac{\varphi}{V_{dA}}}}{\epsilon_s} \tag{5.6}$$

Here ϵ_s is the permittivity of the semiconductor. Noting that $\frac{d\left(\frac{d\varphi}{dx}\right)^2}{dx} = 2\frac{d\varphi}{dx}\frac{d^2\varphi}{dx^2}$, we can modify the above equation as follows

$$\frac{d\left(\frac{d\varphi}{dx}\right)^2}{dx} = 2\frac{qn_{di}e^{\frac{\varphi}{V_{dA}}}}{\epsilon_s}\frac{d\varphi}{dx} \tag{5.7}$$

The above equation can be written in terms of a relation between the electric field, $\xi = -\frac{d\varphi}{dx}$ and the potential φ as

$$d\xi^2 = 2\frac{qn_{di}e^{\frac{\varphi}{V_{dA}}}}{\epsilon_s}d\varphi \tag{5.8}$$

Solving the above differential equation with the condition that $\varphi = 0$ when $\xi = 0$, we find

$$\xi = (2qV_{dA}n_{di}/\epsilon_s)^{1/2}(e^{\frac{\varphi}{V_{dA}}} - 1)^{1/2} \tag{5.9}$$

At $x = 0$, $\varphi = \varphi_s$ where φ_s is the surface potential. Here we can make an assumption that the semiconductor in the thin film transistor is really thin (about 50 nm to 100 nm). On the other hand, the band bending in the semiconductor due to charges trapped in the deep states is linear and therefore we can assume that the electric field is almost constant. The total trapped charge per unit area in the deep states, n_{d-tot} is simply given by $\epsilon_s\xi_{avg}t_s$, i.e.,

$$n_{d-tot} \approx n_{d0}e^{\frac{\varphi_s}{2V_{dA}}} \tag{5.10}$$

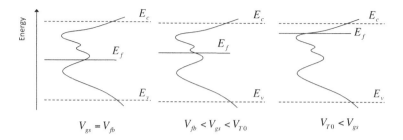

FIGURE 5.1
Fermi level movement in the states for different gate bias. The x-axis represents the density of states.

where $n_{d0} = (2qV_{dA}n_{di}\epsilon_s)^{1/2}t_s$ and t_s is the semiconductor thickness.

After the gate voltage is raised sufficiently large so that all deep state carriers are filled, the Fermi level is close enough to the conduction band edge so as to induce a large number of free carriers. This gate voltage is called the *threshold voltage*. By definition the threshold voltage should be $V_{T0} = V_{fb} + n_{d-tot}/C_i$, where V_{fb} is the flat-band voltage, and C_i is the dielectric capacitance per unit area.

5.2.3   Above Threshold Operation

As the gate bias is increased further, the Fermi level moves into the acceptor-like tail states and the device now operates above the threshold voltage. The charge on the gate is now mirrored by the charges trapped in the deep states and the bandtail states along with the free carrier charge. Thus, Poisson's equation now becomes

$$\frac{d^2\varphi}{dx^2} = \frac{qn_{di}e^{\frac{\varphi}{V_{dA}}} + qn_{ti}e^{\frac{\varphi}{V_{tA}}} + qn_{fi}e^{\frac{\varphi}{V_{th}}}}{\epsilon_s} \tag{5.11}$$

If the density of tail states is very large, we can assume that most of the charge on the gate above the threshold voltage is mirrored by the charge trapped in the tail states. Thus,

$$C_i(V_{gs} - V_{T0}) \approx n_{t-tot} = n_{t0}e^{\frac{\varphi_s}{2V_{tA}}} \tag{5.12}$$

where $n_{t0} = (2qV_{tA}n_{ti}\epsilon_s)^{1/2}t_s$. As the gate bias is pulled up further, the Fermi level attempts to get closer to the conduction band edge. However, the tail state density in non-crystalline semiconductors is large and the Fermi level is pinned in the tail states.

5.3   Transfer Characteristics of the TFT

We now extend the ideas of the MOS capacitor discussed above to define the transfer characteristics of the TFT operating in the above threshold region. The MOS capacitor structure is converted to a TFT by the addition of the source and drain electrodes.

For a given gate-source bias, V_{gs} and a drain to source bias V_{ds}, the charge density along the channel length varies as $C_i(V_{gs} - V_{T0} - V_{ch}(y))$, where $V_{ch}(y)$ is the channel potential, with the channel lying along the y direction between the drain and source. The current through the TFT is given by

$$I_{ds} = \mu q W n_{f-tot} \frac{dV_{ch}}{dy} \qquad (5.13)$$

where n_{f-tot} is the free charge density per unit area, and μ is the mobility of the free electron charge. The free carrier charge per unit area can be related to the trap charge per unit area as

$$n_{f-tot} \approx n_{f0} e^{\frac{\varphi_s}{V_{th}}} = \frac{n_{f0}}{n_{t0}^\alpha}(n_{t0} e^{\frac{\varphi_s}{2V_{tA}}})^\alpha = \frac{n_{f0}}{n_{t0}^\alpha} C_i (V_{gs} - V_{T0} - V_{ch}(y))^\alpha \quad (5.14)$$

where $n_{f0} = n_{fi} t_s$ is the free charge per unit area. Substituting for n_{f-tot} in the equation for current, and integrating along the channel length with y varying from 0 to L, and $V_{ch}(y)$ varying from 0 to V_{ds}, we have the drain to source current in the TFT being

$$I_{ds} = \mu_{eff} \frac{W}{L(\alpha + 1)} C_i^\alpha ((V_{gs} - V_{T0})^{\alpha+1} - (V_{gs} - V_{T0} - V_{ds})^{\alpha+1}) \qquad (5.15)$$

For $\alpha \approx 1$, the current-voltage characterisics for the TFT is the same as for a conventional crystalline silicon MOSFETs, i.e.

$$I_{ds} = \mu_{eff} \frac{W}{L} C_i ((V_{gs} - V_{T0})V_{ds} - \frac{V_{ds}^2}{2}) \qquad (5.16)$$

in the linear region of operation and

$$I_{ds} = \frac{\mu_{eff}}{2} \frac{W}{L} C_i (V_{gs} - V_{T0})^2 \qquad (5.17)$$

in the saturation region of operation. The typical characteristics of a-Si:H based TFTs is shown in Figure 5.2.

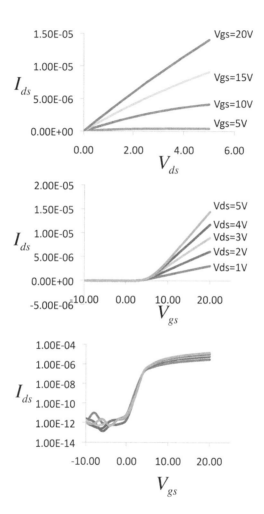

FIGURE 5.2

Typical transfer and output characteristics of a-Si:H TFTs. The channel width is $200\mu m$, and the channel length is $8\mu m$.

5.4   Mobility

In the above discussions, we never really paid much attention to the mobility term in defining the conductivity and therefore the current through the TFT.

In bulk non-crystalline semiconductors, the drift mobility of carriers in the presence of trap states, μ_d is defined by a certain percentage or ratio of the mobility of carriers when there are no trap states, μ_0. This relation is given by

$$\mu_d = \mu_0 \frac{t_{free}}{t_{trap} + t_{free}} \qquad (5.18)$$

where t_{free} is the time spent by the carriers in free states (extended states) and t_{trap} is the time spent by the carriers in the trap states.

In the cases of TFTs we can draw a similar relation. The field effect mobility of the carriers μ_{fe} determines the effective contribution of free and trapped charge to the conductivity, σ, so that $\sigma = \mu_{fe}(n_{f-tot} + n_{t-tot})$. Equivalently, the mobility of the free carriers alone, μ is related to the conductivity as $\sigma = \mu n_{f-tot}$. The relation between the free carrier mobility and the field effect mobility is therefore given by

$$\mu_{fe} = \mu \frac{n_{f-tot}}{n_{f-tot} + n_{t-tot}} \qquad (5.19)$$

Note that in the expression for the current voltage characteristics, the term μ_{eff} contains within it the term $\frac{n_{f0}}{n_{t0}}\mu$ (assuming $\alpha \approx 1$). Since $n_{t0} >> n_{f0}$, the effective mobility is directly related to the field effect mobility in the above threshold region of operation.

5.5   Threshold Voltage Shift

In this section we look at a parasitic effect of charge trapping in non-crystalline semiconductors resulting in a *threshold voltage shift* in field effect transistors based on these semiconductors [107]-[113].

5.5.1   Mechanics of Charge Trapping

When the TFT is operating in the above threshold region, the Fermi level is present in the band tails. The states in the band tails are due to the presence of weak bonds. As these states get filled up, the weak bonds are broken resulting in states deeper in the band gap. This effectively results in the Fermi level slipping down in energy as the carriers fill these newly created deep states. This process of charge trapping in the newly created defect states is a time dependent process resulting in an effective increase in the threshold voltage of the TFT. This mechanism is termed as *defect creation*. The time constants associated with this mechanism are of the order of $1e3$ s to $1e6$ s.

Apart from charge trapping due to defect creation, carriers are also trapped and released from the interfacial states situated at the semiconductor-insulator interface resulting in some additional increase in threshold voltage. This is called *interfacial trapping*. The time constants associated with trapping at the interface states is $1e-6$ s to 1 s.

Charges can also be trapped in deep states within the insulator. This generally happens in poor insulator materials fabricated at low temperatures which have a large number of states close to the semiconductor interface. The electric field results in the thermalization of carriers from the semiconductor-insulator interface to the states inside the insulator. However, for insulators with large band gaps, this mechanism of charge trapping is negligible at typical TFT operating voltages.

While interfacial trapping of carriers is common to both non-crystalline semiconductor and crystalline semiconductors, the mechanism of defect creation is unique to non-crystalline or amorphous semiconductors.

5.5.2 Dynamics of Defect Creation and Threshold Voltage Shift

The dynamics of defect creation is defined by the *stretched exponential function* where $\delta N_d \propto 1 - e^{-(t/t_{0c})^\beta}$. Here δN_d is the created defect state density per unit area, β is a temperature dependent coefficient, and t_{0c} represents a time constant. The creation of defect states with time and results in an effective increase in the threshold voltage of the TFT with time. Since the total charge trapped per unit area due to the created defect states is $q\delta N_d$, the effective threshold voltage shift observed during TFT operation is $\delta V_T = q\delta N_d/C_i$, where δV_T is the threshold voltage shift, and C_i the dielectric capacitance per unit area. For a gate bias stress voltage of V_{gs},

$$\delta V_T = (V_{gs}V_{T0})(1 - e^{-(t/t_{0c})^\beta}) \tag{5.20}$$

where V_{T0} the initial threshold voltage of the TFT. Defining $G(t) = 1 - \delta V_T/(V_{gs} - V_{T0})$, we find

$$\ln(|\ln(G(t))|) = \beta\ln(t) - \beta\ln(t_{0c}) \tag{5.21}$$

The plot of the left hand side of Equation 5.21 as a function of $\ln(t)$ is approximately a straight line with slope β.

Defect creation can be considered to have a certain distribution of activation thresholds with corresponding activation energies. Physically, these activation thresholds correspond to the physio-chemical process of bond breaking and creation of various types of defects. For a TFT under gate bias stress, at a given time t, all potential defect states that have an activation energy less than $kT\ln(\nu_{c0}t)$ will be converted to defects. Here ν_{c0} represents an attempt to escape frequency for defect creation.

The distribution of activation energies for defect creation can be obtained from the dynamics of threshold voltage shift. Deane et. al. plotted δV_T as a function of the energy, E, which is in some sense a cumuluative distribution function of the activation energy. The activation energy distribution was then obtained from the plot of $d(\delta V_T)/dE$ versus E.

5.5.3   Threshold Voltage Recovery

When the gate bias is brought back to flat-band voltage (or generally below the stress voltage), some of the trapped charges are detrapped. While some of the charge is released from the traps at the semiconductor-insulator interface, some of the detrapping occurs due to the removal of defects. Defect removal also has a distribution of activation thresholds such that all defects with an activation threshold less than $kT\ln(\nu_{r0}t)$ are passivated. Here, ν_{r0} is the attempt to escape frequency for defect removal.

5.5.4   Drain-Source Bias Dependence

Since the threshold voltage shift in the TFT is in some sense proportional to the total channel charge (free charge) present at any given gate bias, it must also have a dependence on the drain-source bias.

For a TFT in deep linear mode operation, we can say that the total channel charge is $C_i(V_{gs} - V_{T0})$ and hence the threshold voltage shift is

$$\delta V_{T,lin}(t) = (V_{gs} - V_{T0})f(t) \tag{5.22}$$

where $f(t) = 1 - e^{-(t/t_{0c})^\beta}$ is the stretched exponential function. For a TFT operating in saturation mode of operation, the total channel charge is $2C_i(V_{gs} - V_{T0})/3$ and

$$\delta V_{T,sat}(t) = \frac{2}{3}(V_{gs} - V_{T0})f(t) \tag{5.23}$$

In general from the expression of the total TFT channel charge for any V_{ds}, the general expression for the threshold voltage shift for any V_{gs} and V_{ds} is given by

$$\delta V_T(t) = \frac{2}{3}(V_{gs} - V_{T0})\frac{1 - \phi^3}{1 - \phi^2}f(t) \tag{5.24}$$

where $\phi = 1 - \frac{V_{ds}}{V_{gs} - V_{T0}}$.

5.6 Conclusion

This chapter presented the device analysis of the TFT and its current-voltage characteristics. We found that the states in the gap in the semiconductor causes charge trapping which leads to a threshold voltage shift. Moreover, the asymmetry in the donor-like and acceptor-like states can lead to the non-feasibility of complementary devices i.e., n-type and p-type TFTs. These two features form the essence of the problems of circuit design with the TFT. In the next chapter, we focus our attention on the modeling of the threshold voltage shift.

6

Modeling Threshold Voltage Shift for Circuit Design

CONTENTS

In the previous chapter we looked at the device physics of thin film transistors (TFTs) and briefly studied the threshold voltage shift (V_T shift) in the TFTs. The physics behind the threshold voltage shift involved the mechanisms of defect creation and trapping at the semiconductor insulator interface. The mechanism of defect creation involved the conversion of weak bonds into dangling bonds. This mechanism occurs with different activation energies and therefore the threshold voltage shift in non-crystalline semiconductor based TFTs was shown to have a stretched exponential behavior with time.

In a circuit the TFT experiences a variable gate bias, where the gate-source voltage applied on the TFT switches to different values. From the point of view of the circuit design, a simpler understanding of the dynamics of the threshold voltage shift in the TFT is needed preferably through a compact black-box model [127], [128].

In this chapter we look at some simple linear models for the dynamics of the threshol voltage shift when the TFT is driven by a variable gate-source voltage. We attempt to develop compact models which approximately describe the behavior of the threshold voltage shift with time. These models will be useful in analyzing many of the behavioral properties of the circuits.

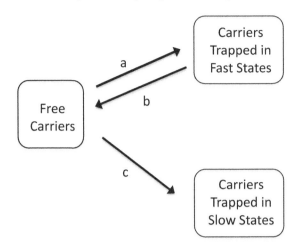

FIGURE 6.1
Linear model for charge trapping.

6.1   Constant Gate Bias

Without defning the physics of the process of charge trappig, we can consider a simple linear model for charge trapping as shown in Figure 6.1. The threshold voltage in the TFT can be seen to be the result of charge trapping in *fast states* and *slow states*. Charges trapped in the fast states are recoverable by removing the gate bias, while the charges trapped in the slow states can be recovered only by annealing the device.

According to Figure 6.1 the time dependence of threshold voltage for a constant gate bias can be shown to be of the form,

$$V_T(t) - V_{T0} = (V_{gs} - V_{T0})(1 - \varphi e^{-\kappa t} - (1 - \varphi)e^{-\omega t}) \qquad (6.1)$$

Here κ is the effective time constant corresponding to the fast states, while ω is the time constant corresponding to the slow states. All charge induced due to field effect will eventually be trapped in either the fast or slow states, with the threshold voltage shift at $t = \infty$ becoming $V_{gs} - V_{T0}$. The total trapped charge at $t = \infty$ is therefore $C_i(V_{gs} - V_{T0})$, where C_i is the capacitance per unit area. φ is a coefficient which tell us the ratio of charge trapped in the fast states to the total trapped charge when $t = \infty$. A more detailed analysis of this system is provided in the Appendix at the end of the book. Figure 6.2 illustrates the model plotted against experiemental data for a a-Si:H TFT. In order to extract the parameters κ, ω, and φ, we follow this simple procedure. The threshold voltage shift is plotted as a function of time, with the time scale atleast comparable to $1/\omega$. We consider the experimental data during

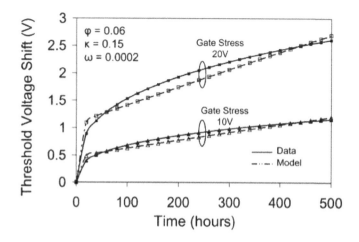

FIGURE 6.2
Model plotted against experimental data for a constant gate bias. (From Sambandan, S.; Lei Zhu; Striakhilev, D.; Servati, P.; Nathan, A.; Markov model for threshold-voltage shift in amorphous silicon TFTs for variable gate bias, IEEE Electron Device Letters, IEEE, Volume: 26, Issue: 6, Digital Object Identifier: 10.1109/LED.2005.848116 Publication Year: 2005, 375–377. With permission.)

till the knee of the curve and draw the average tangent to this part of the curve. This part of the curve represents the faster time constants and the rapid charge trapping in the fast states. The slope of this tangent is indicative of the time constant κ^{-1}. Similarly we draw a tangent to the curve beyond the knee. This tangent represents the slower states seen within the time frame of the experiment. The slope of this tangent is indicative of the time constant ω^{-1}. If the threshold voltage shift till the knee of the curve is V_{T-knee}, then $\varphi = V_{T-knee}/(V_{gs} - V_{T0})$.

We now consider the case when the gate bias is constant for a short period of time ($\approx \kappa^{-1} << \omega^{-1}$). In such a case, the short-term time dependence of the threshold voltage of the TFT under constant gate bias is given by

$$V_T(t) - V_{T0} \approx \varphi(V_{gs} - V_{T0})(1 - e^{-\kappa t}) \tag{6.2}$$

The above approximation for the dynamics of short-term threshold voltage shift is equivalent to the dynamics of charging a capacitor of an RC circuit to a voltage of $\varphi(V_{gs} - V_{T0})$ with the time constant of the RC circuit being κ.

6.2   Removal of Gate Bias

When the gate bias to the TFT is removed ($V_{gs} = 0$) after a short stress period, the threshold voltage begins to recover. This recovery of the threshold voltage is akin to the dynamics of the RC capacitor during the discharge of the capacitor.

For example, let the TFT be first stressed with a constant gate bias of V_{gs1} for a certain time period. At the end of this time, the threshold voltage of the TFT is $V_{T,1}$, which is greater than the initial threshold voltage V_{T0}. Then the gate bias of the TFT is removed, and the TFT relaxes with the threshold voltage returning to its original value of V_{T0}.

Using the capacitor discharge model, the dynamics of the recovery of the threshold voltage shift after removal of the gate bias can thus be defined as

$$V_T(t) - V_{T0} \approx (V_{T,1} - V_{T0})e^{-\kappa t} \qquad (6.3)$$

Here we have reset the time zero to begin when the relaxation starts.

6.3   Variable Gate Bias

We now try to understand what happens to the threshold voltage shift in the TFT when the gate bias to the TFT is varied.

Let us consider a TFT that is first stressed at a gate bias, V_{gs1}, and then stressed at another gate bias V_{gs2}. The change in gate bias of the TFT can be considered to be an independent combination or superposition of — first, the removal of V_{gs1} leading to a threshold voltage recovery, and second, the application of V_{gs2} leading to a threshold voltage increase.

If the threshold voltage at the end of the stress by V_{gs1} is $V_{T,1}$, the threshold voltage dynamics during the application of bias V_{gs2} can be writen as,

$$V_{T,2} - V_{T0} = (V_{T,1} - V_{T0})e^{-\kappa t} + \varphi(V_{gs2} - V_{T0})(1 - e^{-\kappa t}) \qquad (6.4)$$

Here, the term $(V_{T,1} - V_{T0})e^{-\kappa t}$ corresponds to the recovery process, while the term $\varphi(V_{gs2} - V_{T0})(1 - e^{-\kappa t})$ corresponds to the charge trapping process.

6.3.1   Generalizations

In order to generalize the above results, we now divide the time-line into equal intervals of length τ, with the gate bias changing in each interval. The gate bias applied to the TFT in the time interval $(j - 1)\tau \leq t < j\tau$ is $V_{gs,j}$. If the gate bias is constant in two subsequent time intervals it simply implies $V_{gs,j-1} = V_{gs,j}$.

Using the previously discussed ideas, the threshold voltage at time $t = n\tau$ is given by

$$V_{T,n} - V_{T0} = (V_{T,n-1} - V_{T0})e^{-\kappa\tau} + \varphi(V_{gs,n} - V_{T0})(1 - e^{-\kappa\tau}) \quad (6.5)$$

This recursive relation can be expanded by writing out the expression for $V_{T,n-1}$ and subsequently all previous values of threshold voltage to yield

$$V_{T,n} = V_{T0} + \varphi(1 - e^{-\kappa\tau}) \sum_{j=1}^{n} (V_{gs,j} - V_{T0})(e^{-\kappa\tau})^{n-j} \quad (6.6)$$

This is a very useful result in terms of circuit applications. Figure 6.3 illustrates experimental data with model fits.

6.3.2 Thought Experiments

6.3.2.1 First Thought Experiment

Our first experiment verifies if Equation 6.6 holds its integrity when the gate bias is not really varied. If the TFT is stressed with a constant gate bias from $t = 0$ to $t = T$ with a constant gate bias, V_{gs0}, the threshold voltage shift according to Equation 6.2 is given by

$$V_{T,n} = V_{T0} + \varphi(V_{gs0} - V_{T0})(1 - e^{-\kappa T}) \quad (6.7)$$

We should be able to obtain the same result if we assume that the time interval is divided in small segments each of length τ such that $T = n\tau$ and that the same V_{gs0} is the bias to the TFT in every time segment. The threshold voltage shift under this case as predicted by Equation 6.6 is

$$
\begin{aligned}
V_{T,n} &= V_{T0} + \varphi(V_{gs0} - V_{T0})(1 - e^{-\kappa\tau}) \sum_{j=1}^{n} (e^{-\kappa\tau})^{n-j} \quad (6.8) \\
&= V_{T0} + \varphi(V_{gs0} - V_{T0})(1 - e^{-\kappa\tau}) \frac{1 - e^{-n\kappa\tau}}{1 - e^{-\kappa\tau}} \\
&= V_{T0} + \varphi(V_{gs0} - V_{T0})(1 - e^{-\kappa T})
\end{aligned}
$$

This shows consistency in the model described by Equation 6.6.

6.3.2.2 Second Thought Experiment

We now perform another thought experiment that checks for the validity of the function used for the dynamics, i.e., is the function $1 - e^{-\kappa t}$ a valid representation of the threshold voltage shift dynamics? Let us pick any random function $f(t)$ to represent the dynamics of the threshold voltage shift. There are some elementary properties $f(t)$ must have. Firstly, $f(t = \infty) = 1$ and $f(t = 0) = 0$. Secondly, the function describing the threshold voltage shift

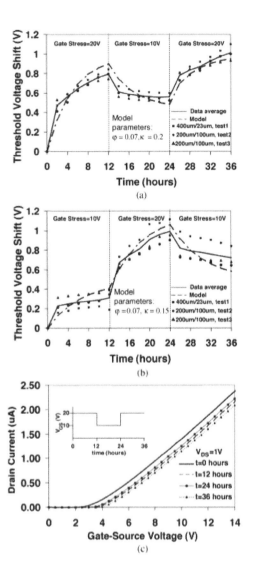

FIGURE 6.3
Model plotted against experimental data for a time varying gate bias. (From Sambandan, S.; Lei Zhu; Striakhilev, D.; Servati, P.; Nathan, A.; Markov model for threshold-voltage shift in amorphous silicon TFTs for variable gate bias, IEEE Electron Device Letters, IEEE, Volume: 26, Issue: 6 Digital Object Identifier: 10.1109/LED.2005.848116 Publication Year: 2005, 375–377. With permission.)

should have a deacceleration property, which implies that the threshold voltage shift with time must slow down as there is less tendency for the charge to get trapped with time. Hence, $d^2 f(t)/dt^2 < 0$. This implies that the general nature of Equation 6.2 and Equation 6.3 remain the same with the exception that the dynamics is defined by $f(t)$ instead of $1 - e^{-\kappa t}$. Thus, the modified form of Equation 6.2 for the short time threshold voltage shift under a bias of V_{gs} becomes

$$V_T(t) - V_{T0} \approx \varphi(V_{gs} - V_{T0})f(t) \tag{6.9}$$

and the modified form of Equation 6.3 for the short time threshold voltage recovery becomes

$$V_T(t) - V_{T0} = (V_{T,1} - V_{T0})(1 - f(t)) \tag{6.10}$$

Therefore, if the TFT is stressed with a constant gate bias V_{gs0} for a time $t = T_j$, the threshold voltage at the end of the stress period is defined by

$$V_{T,T_j} - V_{T0} = \varphi(V_{gs0} - V_{T0})f(T_j) \tag{6.11}$$

Now let us imagine that the TFT is stressed with a constant gate bias V_{gs0} with an imaginary stop at some time T_i where $0 < T_i < T_j$. The threshold voltage shift at $t = T_i$ is given by Equation 6.5 to be

$$V_{T,T_i} - V_{T0} = \varphi(V_{gs0} - V_{T0})f(T_i) \tag{6.12}$$

Then, proceeding from time $t = T_i$ to $t = T_j$ with the same gate bias, the threshold voltage according to the model Equation 6.5 must predict

$$V_{T,T_j} - V_{T0} = (V_{T,T_i} - V_{T0})(1 - f(T_j - T_i)) + \varphi(V_{gs0} - V_{T0})f(T_j - T_i) \tag{6.13}$$

Substituting for V_{T,T_i} from Equation 6.12 in Equation 6.13, we obtain

$$V_{T,T_j} - V_{T0} = \varphi(V_{gs0} - V_{T0})[f(T_i + f(T_j - T_i) - f(T_i)f(T_j - T_i)] \tag{6.14}$$

The result predicted by this model, Equation 6.14 using the imaginary time stop must be same as the result predicted for the threshold voltage of the TFT without any imaginary time stop, Equation 6.11. Equating the two results

$$\varphi(V_{gs0} - V_{T0})f(T_j) = \varphi(V_{gs0} - V_{T0})[f(T_i + f(T_j - T_i) - f(T_i)f(T_j - T_i)] \tag{6.15}$$

Substituting $g(t) = 1 - f(t)$, Equation 6.15 implies

$$g(T_j) = g(T_i)g(T_j - T_i) \tag{6.16}$$

Clearly, $g(t) = e^{-\kappa t}$ is a valid function and $f(t) = 1 - e^{-\kappa t}$ is a valid representation of the dynamics of the threshold voltage shift.

6.4   Conclusion

This chapter presented a recursive model for defining the threshold voltage of the TFT for variable gate bias conditions. This model plays a useful role when we develop circuit design techniques for the TFT.

Part III

Thin Film Transistor Circuits and Applications

7

Transistor as a Switch

CONTENTS

An electronic switch is perhaps one of the most fundamental of circuit elements. Consider the situation shown in Figure 7.1 where an electrical signal at point A is to be transmitted to point B only at certain instances in time. This is achieved by having a conducting line between point A and point B with a switch in between. When the switch is closed, any variation in the signal at point A is also observed at point B with allowances for the loss along the conductor and the switch. When the switch is open, signal variations at point A will not be observed at point B, again with allowances for the resistance of the gap in the conductor.

From this description, the features of an ideal switch become apparent. Some of the parameters that determine the figure of merit for an electrical switch are as follows,

- *On resitance* — The on resistance is the resistance of the switch when it is closed. The on resistance of an ideal switch is zero. Moreover, an ideal switch must not have any signal delay or energy storage — inductive or capacitive elements — when it is on. When the switch is closed, it must faithfully and immediately pass on the signal observed at one end to the other.

- *Short circuit current capacity* — The short circuit current capacity is the maximum amount of current a switch can transmit without destroying itself when closed. The current capacity of an ideal switch is infinite.

- *Off resistance* — The off resistance is the resistance of the switch when it is open. The off resistance of an ideal switch is infinite.

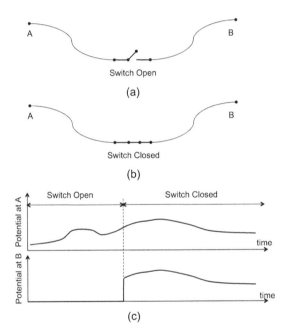

FIGURE 7.1
Concept of a switch. (a) The open switch isolates point A from point B electrically (b) The closed switch connects point A and point B electrically (c) Voltage fluctuations at point A are observed at point B when the switch is closed.

- *Open circuit voltage capacity* — The open circuit voltage capacity is the maximum voltage a switch can hold across its terminals without destroying itself when open. The open circuit voltage capacity of an ideal switch is infinite.

- *Switching time* — The switching time defines the time required to transit from on to off or vice versa after it has been signalled to do so. The ideal switch will have a zero switching time.

- *Switching dynamics* — The switching dynamics define the dynamics of the switch immediately after transition from on to off or vice versa. If the switch is poorly damped, the switch will bounce between open and closed positions before settling. An ideal switch will have no state transient after it switches.

A thin film transistor (TFT) can be used as an electrical switch. The gate terminal is used to control the switch, i.e., keep it open or close. When the switch is closed, a conductive channel of carriers connect the source to the drain. When the switch is open, the absence of the channel of carriers disconnect the source from the drain electrode. In this chapter we consider the use of the TFT as a switch.

7.1 Transistor Biasing for Switch Operation

Figure 7.2 describes how the TFT switch replaces the switch of Figure 7.1. The drain and source terminals connect points A and B while the gate terminal receives a control signal that commands the switch to either turn on (close) or turn off (open). In this particular example we use an n-channel TFT.

When the switch is open, very little current must flow through the switch, and hence the gate-source voltage must be less than the threshold voltage of the TFT. This will turn off the switch. In order to turn on (close) the switch, the absolute gate-source voltage, $|V_g - V_s|$, must be greater then the absolute threshold voltage, $|V_T|$. But there is one further condition. Since, an ideal switch must transmit the variations in the signal at point A to point B faithfully, it should not be biased in saturation. As observed in the output characteristics of the n-type and p-type TFTs (Figure 7.3), variations in the drain-source voltage do not change the current much, and hence the TFT has a high impedance when biased in saturation. Thus, for operation as a switch, the TFT must always be biased in linear operation.

Figure 7.4 illustrates the operation of the n-channel TFT switch shown in Figure 7.4a. When the switch is biased in linear mode as shown in Figure 7.4b, the potential at point B follows the potential at point A faithfully when the switch is closed. On the other hand, when the switch is biased in saturation

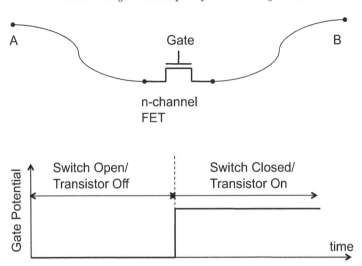

FIGURE 7.2
A field effect transistor used as an electrical switch. The gate terminal is used to control the switch, i.e., keep it open or closed.

mode of operation Figure 7.4c, the potential at point B does not follow the fluctuation at point A faithfully due to the high impedance of the switch.

7.2   On Resistance

When the TFT is used as a switch, it is biased in linear operation where $I_{ds} = \beta(V_{gs} - V_T - V_{ds}/2)V_{ds}$, where $\beta = \mu C_{ox} W/L$. Consider the circuit where the TFT switch is in series with a load capacitor, C. A data voltage of V_d is present at the drain terminal of the switch. The switch is initially open (transistor is off with infinite off resistance), and the voltage on the capacitor is at ground potential. After the switch is closed, the capacitor begins to charge with $v_c(t)$ being the potential across the capacitor at some intermediate time. Note the gate-source voltage across the switch is $V_g - v_c(t)$, and the drain-source voltage is $V_d - v_c(t)$. Here, V_g and V_d imply gate-ground and drain-ground voltage. The equation governing the dynamics of charging

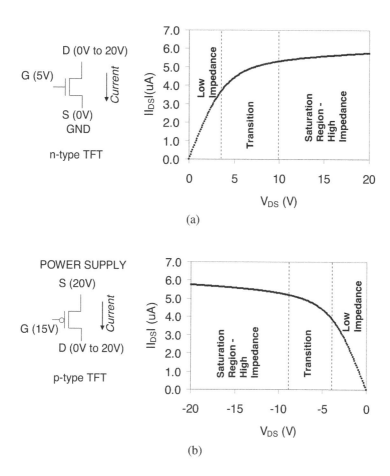

FIGURE 7.3

(a) Output characteristics of the n-channel TFT. (b)Output characteristics of the p-channel TFT. For close to ideal operation, a switch when closed must be biased in above threshold linear region.

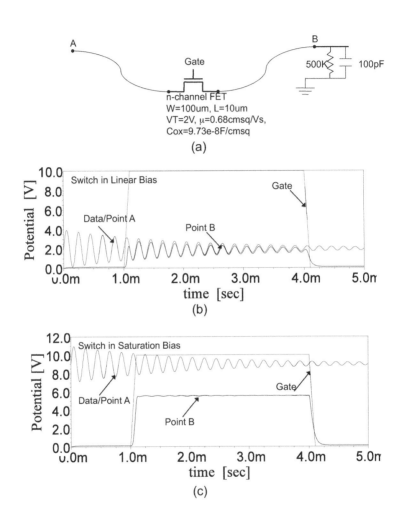

FIGURE 7.4
Simulation of the influence of TFT bias on switch performance. (a) The circuit simulated. (b) Switch biased in linear mode of operation. (c) Switch biased in saturation mode of operation.

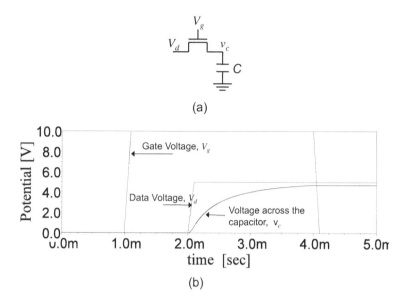

(a)

(b)

FIGURE 7.5
(a)Switch capacitor circuit where the data voltage, V_d, is to be written onto the capacitor via an acess TFT switch. (b) AIM-SPICE simulations of the circuit illustrating the dynamics of capacitor charging.

of the capacitor is

$$C\frac{dv_c}{dt} = I_{switch} \tag{7.1}$$
$$= \beta[(V_g - v_c(t)) - V_T - (V_d - v_c(t))/2][V_d - v_c(t)]$$
$$= \frac{\beta}{2}[V_b - v_c(t)][V_d - v_c(t)]$$

where $V_b = 2(V_g - V_T) - V_d$. Note that $V_b > V_d$ since the TFT is biased in linear mode of operation. Solving this we find that the charge on the capacitor at any given time, t is defined as,

$$v_c(t) = \frac{V_d(1 - e^{-\frac{\beta}{2C}(V_b - V_d)t})}{1 - \frac{V_d}{V_b}e^{-\frac{\beta}{2C}(V_b - V_d)t}} \tag{7.2}$$

When $t = 0$, $v_c = 0$, and when $t = \infty$, $v_c = V_d$ as expected. The time constant of the circuit is $\approx \frac{2C}{\beta(V_b - V_d)} = \frac{C}{\beta(V_g - V_T - V_d)}$. Figure 7.5 illustrates the dynamics of the model of Equation 7.2 using AIM-SPICE simulations.

Small Signal On Resistance

It seems that there is another way to derive the dynamics of the switch capacitor circuit. Since the switch and capacitor make an RC circuit, we can say

$$C\frac{dv_c}{dt} = (V_d - v_c(t))/R_{switch} \tag{7.3}$$

The on resitance of the switch for small variations in V_{ds} is

$$\frac{\partial V_{ds}}{\partial I_{ds}} = \frac{1}{\beta(V_{gs} - V_T - V_{ds})} \tag{7.4}$$

We call this the *small signal on resistance* of the TFT. Using this value of the resistance in place of R_{switch} in Equation 7.3, we can write the dynamics of the switch as

$$\begin{aligned} C\frac{dv_c}{dt} &= \beta[V_d - v_c(t)][(V_g - v_c(t)) - V_T - (V_d - v_c(t))] \tag{7.5}\\ &= \beta V_n[V_d - v_c(t)]. \end{aligned}$$

Here $V_n = V_g - V_T - V_d$. We find the dynamics to be defined by the following equation

$$v_c(t) = V_d(1 - e^{-\frac{\beta}{C}V_n t}) \tag{7.6}$$

When $t = 0$, $v_c = 0$, and when $t = \infty$, $v_c = V_d$ as expected. The time constant associated with the capacitor charging is $\frac{C}{\beta V_n} = \frac{C}{\beta(V_g - V_T - V_d)}$.

However, the dynamics defined by Equation 7.6 is clearly different from that described by Equation 7.2. What may be the reason for this discrepancy? The reason is that by writing Equation 7.3, we have implicitly assumed the TFT is a linear device, i.e., it obeys Ohm's law. This is true for small variations in V_d, thereby permitting us to linearize the current-voltage characteristics and define a resistance as in Equation 7.4. If we make the assumption that V_d is very small, $V_d << V_b$ and the denominator of Equation 7.2 becomes unity. Noting that $V_b - V_d = 2V_n$, Equation 7.2 becomes Equation 7.6. However, in general Equation 7.2 is more correct than Equation 7.6. The difference in the two assumptions is illustrated in Figure 7.6 where the models for the TFT switch dynamics as described by Equation 7.2 and Equation 7.6 are plotted against the AIM-SPICE simulations described in Figure 7.6. We see that Equation 7.2 is more accurate in describing the dynamics as compared to the small signal

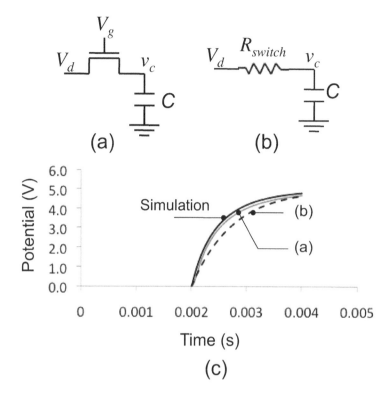

FIGURE 7.6
(a) Equivalent circuit of the switch-capacitor using the TFT current in linear operation. (b) Approximation of the TFT switch with an equivalent resistance. (c) Comparison of models of capacitor charging using the circuit of (a) and (b) with comparisons to AIM-SPICE simulation of Figure 7.5

model of Equation 7.6. However, the small signal approximation has its advantages since it relates to the simple RC circuit, and makes it easy to identify the time contant of the dynamics. This is generally useful for quick calculations in circuit design where the designer chooses the aspect ratio of the TFT to achieve the time constants desired.

7.3   Off Resistance

Ideally the off resistance of the switch is infinite. However, the TFT switch when turned off will have some finite leakage current through the source and drain. If the leakage is too large, it can cause serious signal degradation.

Consider a TFT switch-capacitor circuit with the capacitance of the capacitor being C_0 and the initial voltage on the capacitor being V_i. If the drain terminal of the switch is at voltage V_0, the voltage drop across the TFT is $V_{ds} = V_0 - V_i$. When the TFT is off $(V_{gs} - V_T << 0)$, this voltage drop corresponds to a leakage current I_l through the TFT, and this leakage current is dependent on V_{gs} and V_{ds}. If we assume that I_l is constant, the voltage on the capacitor after time T is $V_f = V_i + I_l T/C_0$ if $V_0 > V_i$ and $V_f = V_i - I_l T/C_0$ if $V_0 < V_i$. If I_l can be written as $V_{ds}\Theta$, with Θ being a constant, we have

$$V_f = V_i + T(V_0 - V_i)\Theta/C_0 \qquad (7.7)$$

This degradation of the original signal on the capacitor, V_i, can be significant for large I_l.

The transfer characteristics of an a-Si:H TFT is shown in Figure 7.7. The characteristics can be divided into four regions — the above threshold region, the forward subthreshold region, the reverse subthreshold region, and the Poole Frenkel region. The above threshold and forward subthreshold region of operation correspond to the on state of the TFT. Once the TFT is switched off, depending on the gate bias, the TFT operates in either the reverse subthreshold region or the Poole Frenkel region. In this section our interest lies in studying the mechanisms of leakage in the TFT.

The main conduction in the reverse subthreshold region is along the *back channel* of the TFT. The back channel implies the formation of a channel away from the semiconductor-insulator interface near the gate. If the TFT has an insulator encapsulation, a weak channel of electrons is formed at the back semiconductor-insulator interface. If the drain bias is positive, there appears a non-zero leakage current along this interface. This current is exponentially dependent on the gate-source bias and the drain bias and is defined as

$$I_{L-rs} \propto I_{0-rs}(W/L)e^{V_{gs}/G_r v_{th}}e^{|V_{ds}|} \qquad (7.8)$$

where, $G_r v_{th}$ is the reverse subthreshold slope, and I_{0-rs} is the reverse subthreshold leakage current prefactor.

If the a-Si:H TFT is switched off at an even lower gate bias, we have hole accumulation at the *front semiconductor-insulator interface*. The front interface refers to the semiconductor-insulator interface close to the gate. At high positive drain voltages, there appears a significantly high leakage current. This current is due to holes generated by Poole-Frenkel emission at the drain-gate overlap area. This leakage current is defined as

$$I_{L-pf} \propto I_{0-pf}WL_{ov}e^{V_{gs}/\alpha_{pf}}e^{|V_{ds}|} \qquad (7.9)$$

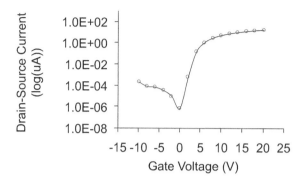

FIGURE 7.7
Transfer characteristics of an a-Si:H TFT.

where, L_{ov} is the drain/source-gate overlap length, α_{pf} the Poole Frenkel parameter, and I_{0-pf}, the Poole Frenkel leakage current prefactor.

7.4 Switching Time

The switching time of the TFT is defined by how quickly the TFT current can move from the off current level to the on current level when the gate voltage switches from the off level below the threshold voltage to the on level above the threshold voltage. Crystalline semiconductor based MOSFETs have a short switching time since the Fermi level is free to move through the empty band gap. However, the non-crystalline semiconductor based TFTs have a more gradual transition regime with the transition becoming more gradual (slow) as we move from crystalline to glassy/amorphous semiconductors.

The region from flat-band to the inversion region is the subthreshold region of operation. In the subthreshold region of operation the drain current is exponentially dependent on the gate-source bias and drain-source bias so that,

$$I_{ds} = I_{ds0}e^{(V_{gs}-V_{fb})/Gv_{th}}\left(1 - e^{-V_{ds}/Gv_{th}}\right) \qquad (7.10)$$

An important measure of how fast the switch turns on is the *subthreshold slope*. From the plot of $log_{10}(I_{ds})$ vs V_{gs} we focus on the subthreshold regios which appears linear due to the exponential dependence of I_{ds} on V_{gs}. The slope measured in V/decade of current is the subthreshold slope.

It is clearly seen that the subthreshold slope is Gv_{th}. The parameter G is the one that determines how fast the TFT turns on.

$$G = dV_{gs}/d\varphi_s. \tag{7.11}$$

Here φ_s is the surface potential.

$$\varphi_s \;=\; V_{gs} - V_{ox} \tag{7.12}$$

$$V_{ox}C_{ox} \;=\; Q_i + Q_s \tag{7.13}$$

where Q_i is the charge trapped in the interface states, and Q_s is the charge trapped in the deep states of the semiconductor. We define an effective interface and deep state capacitance, C_i and C_s so that, $Q_i = \varphi_s C_i$, and $Q_s = \varphi_s C_s$. From Equation 7.11, Equation 7.13, and the above definitions, we find

$$G = 1 + \frac{C_i}{C_{ox}} + \frac{C_s}{C_{ox}} \tag{7.14}$$

7.5   Parasitics

7.5.1   Threshold Voltage Shift

We have seen earlier that non-crystalline semiconductor based TFTs experience a threshold voltage shift in time with applied gate bias. This shift was attributed to charge trapping in the insulator as well as the creation of defect states in the semiconductor. The shift in threshold voltage makes the on resistance of the TFT a function of time.

The threshold voltage shift, δV_T in the TFT was derived to be linearly dependent on the gate bias, while also being a function of the drain-source bias. The time dependence of the threshold voltage shift is defined by the stretched exponential function such that,

$$\delta V_T = \frac{2}{3}(V_{gs} - V_{T0})\frac{1 - \varphi^3}{1 - \varphi^2}(1 - exp(-(t/t_{0c})^\beta)) \tag{7.15}$$

where, t_{0c} is a time constant, β a temperature dependent coefficienct, $\varphi \approx 1 - V_{ds}/V_{gs}$, and V_{T0} the initial threshold voltage. Since a switch operates in linear regime, we find that $0 < \varphi < 1$. If we make the assumption that the switch is biased in deep linear mode, where $V_{ds} << V_{gs}$, it was shown that

$$\delta V_T = (V_{gs} - V_{T0})(1 - exp(-(t/t_{0c})^\beta)) \tag{7.16}$$

The small signal on resistance of the FET defined in Equation 7.4 is modified to accommodate the impact of threshold voltage shift, so that

$$R_{switch} = \frac{\partial V_{ds}}{\partial I_{ds}} = \frac{1}{\beta((V_{gs} - V_{T0})e^{-(t/t_{0c})^\beta} - V_{ds})} \tag{7.17}$$

The VT shift in the TFT increases the on resistance, and therefore the time constant of the dynamics of capacitor charging in a switch-capacitor circuit. However, the time dependence of the VT shift need not be considered while solving for the dynamics of capacitor charging in Equation 7.2 since the time constants associated with the VT shift and capacitor charging are very different. VT shift being a typically slow phenomena has a much larger time constant compared to the dynamics of capacitor charging in a switch-capacitor circuit.

7.5.2 Clock Feedthrough

The *clock feedthrough* refers to the charge feeding through the overlap between the gate and source/drain when the gate signal changes [129]. When a TFT switch is used to access a capacitor, a voltage data is written on the capacitor by first closing the switch. After the data is written, the switch is opened in order to save the data. This opening of the switch involves the gate signal to the TFT making a high ($> V_T$) to low ($< V_T$) transition in the case of a n-type TFT. This transition of the gate voltage feeds charge to the drain and source electrodes and hence the capacitor due to clock feedthrough. This corrupts the data stored on the storage capacitor. The error in data voltage is given by

$$v_{e_cf} = (V_{Gon} - V_{Goff})\frac{WLC_{ox}}{WLC_{ox} + C_L} \qquad (7.18)$$

This error is independent of data voltage and is just an offset.

7.5.3 Charge Injection

Charge injection is another mechanism that corrupts the data on a capacitor in a TFT switch-capacitor circuit [129]. Once again we imagine a TFT switch-capacitor circuit where a voltage data is written on the capacitor by first closing the switch. If the switch is turned off (opened) very quickly, a part of the charge in the channel is dumped onto the capacitor. This charge corrupts the written data. The error due to charge injection is given by,

$$v_{e_ci} = [\frac{[C_{line} - C_L]}{2C_L}\frac{r}{r_o} + 1]\frac{1}{C_{line} + C_L}Q_{ch} \qquad (7.19)$$

Here r is the rate at which the gate line is turned on or off, and r_o is a rate coefficient and is intuitively the rate at which C_L will discharge a small amount of charge through the switch. This is a more serious error as it depends on the data voltage.

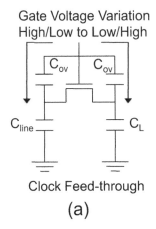

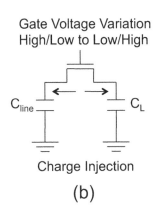

FIGURE 7.8
(a) Mechanism of clock feedthrough where the time-varying gate voltage influences the charge on the load capacitor via coupling through the overlap capacitance. (b) Mechanism of charge injection where the time varying gate voltage pumps the channel charge in the TFT channel to the external capacitors.

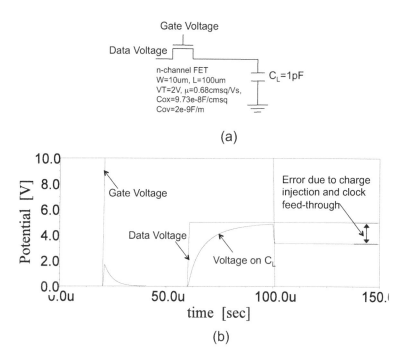

FIGURE 7.9

(a) Circuit used for simulating the effects of charge injection and clock feedthrough. (b) AIM-SPICE simulations of charge injection and clock feedthrough. A rapidly transiting gate voltage induces error due to the mechanisms of charge injection and clock feedthrough.

7.6   Conclusion

This chapter presented the concepts related to the use of TFTs as a switch. The switch is a fundamental circuit unit and the ideas presented in this chapter are preliminary to the design of digital circuits. The switch-capacitor circuit is also a fundamental circuit which is the primary circuit element in most large area electronic systems.

8

Diode Connected Transistor

CONTENTS

This chapter is dedicated to a unique configuration of the field effect transistor — *a diode connected transistor*. The ideal conventional p-n junction diode is shown in Figure 8.1a. Ideally, if the potential at point A, V_A, is greater than the potential at point B, V_B, such that $V_A - V_B > V_T$, the threshold voltage of the diode, there is a flow of current from A to B. On the other hand, if $V_A - V_B < V_T$, there is no current from A to B as illustrated by Figure 8.1d. In other words the diode can be considered to be a circuit element which permits current flow in just one direction — from the p-side to the n-side — if the potential drop from p to n is greater than the threshold voltage of the diode. On the other hand, if the diode is reverse biased with the potential at point B being greater than the potential at point A, no current flow, except for a small leakage current, is allowed. This is exactly what a diode connected transistor achieves. This configuration is given importance because of its particular usefulness as a biasing element in large area electronic systems and is the main focus of this chapter.

8.1   Circuit Configuration and Operation

The circuit of a diode connected n-channel and p-channel TFTs are shown in Figure 8.1a and Figure 8.1b, respectively. The diode configuration is defined by the gate of the TFT connected to the drain terminal. Since the gate-source and drain-source potential are the same, the TFT is always in saturation bias.

     Consider the n-channel TFT. We label two terminals, the drain of the TFT as point A, and the source of the TFT as point B. When V_A is greater

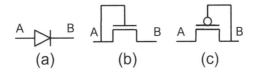

(a)   (b)   (c)

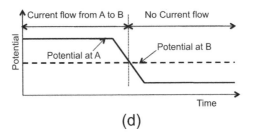

(d)

FIGURE 8.1

(a) Conventional diode (b) Diode connected n-channel TFT (c) Diode connected p-channel TFT.

than V_B, such that $V_A - V_B$ is greater than the threshold voltage of the TFT, V_T, we have a current flow from point A to point B. On the other hand, if $V_A - V_B < V_T$ there is no flow of current except for the leakage. In other words, the circuit of Figure 8.1b allows a flow of current from A to B if the potential at point A exceeds the potential at point B by the threshold voltage of the TFT. This configuration of the TFT performs the same function of the diode of Figure 8.1a in this limited sense as illustrated by Figure 8.1d.

Here, it must be noted that even though the transistor is in saturation bias, it must not be confused with having a high impedance in this case. This is because the gate voltage of the TFT varies with variations of the drain voltage (since $V_{gs} = V_{ds}$), therefore making the resistance of the TFT very sensitive to the drain-source voltage. For a drain-source voltage $V_{ds} > V_T$, the current through the diode connected TFT is defined as

$$I_d \approx \beta(V_{ds} - V_T)^2 \tag{8.1}$$

Here $\beta = \mu C_i W / 2L$, where μ is the mobility, C_i the gate capacitance per unit area, W the channel width, and L the channel length. The effective *forward bias small signal resistance* of the diode connected TFT is therefore given by

$$R_d = \frac{\partial V_{ds}}{\partial I_d} = \frac{1}{2\beta(V_{ds} - V_T)} \tag{8.2}$$

An AIM-SPICE simulation of the operation and dynamics of the diode connected n-channel TFT charging a capacitor is shown in Figure 8.2. We see that as long as the applied data voltage, V_{ds} is greater than the voltage on the

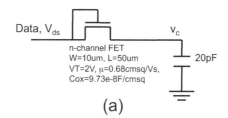

(a)

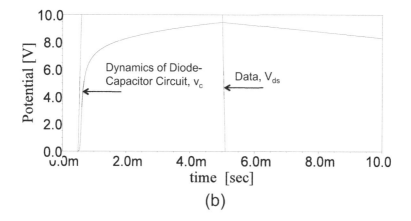

(b)

FIGURE 8.2
(a) TFT diode-capacitor series circuit. (b) Simulation of the dynamics of the voltage on the capacitor to a pulse input voltage. The diode configuration enables the capacitor to hold the input peak after the pulse input has passed. There are losses due to the diode leakage.

capacitor plate, v_c, there is a current flow charging the capacitor. However, once the capacitor voltage is greater than the data voltage, there is no current flow except for the leakage through the TFT.

8.2 Applications

We introduced the diode connected TFT by saying that this configuration was particularly useful for large area electronic systems. In this section we list out some specific areas of application of such a configuration.

8.2.1   Circuit Biasing

Connecting a transistor in this fashion is a useful configuration to ensure that certain transistors in a circuit always operate in saturation irrespective of the drain potential as long it is above the threshold voltage. This is therefore a useful tool for biasing cicuits correctly.

The diode connected TFT can be thought of as a two terminal device since the drain is connected to the gate. Moreover, the current through the TFT is entirely decided by the potential drop across these two terminals. Equivalently, if the current through the diode connected TFT is fixed, the potential across the two terminals is unique. It is easy to see from Equation 8.1 that

$$V_{ds} = \left(\frac{I_d}{\beta}\right)^{1/2} + V_T \qquad (8.3)$$

This is useful in providing several bias voltage levels for circuits as illustrated in Figure 8.3. Consider the case where the circuit requires several different bias voltage levels. One option is to provide several bias lines as shown in Figure 8.3a. However, this is expensive in terms of layout area, more so if the circuit is arrayed as in the case of most large area electronic systems. Here, the diode connected TFT provides a solution as shown in Figure 8.3b. With a single current source, a series of diode connected TFTs of suitable aspect ratios, can provide all the bias signals needed for the circuit. The key however, is to ensure that the aspect ratios of the diode connected TFTs are large enough so as to provide the current levels required for the circuit and minimize the RC delays. If the bias voltages are provided to capacitive nodes of the circuit, e.g., the gate of TFTs, the dc current requirements are negligible. Figure 8.3c shows AIM-SPICE simulations of the dynamics of the circuit while providing the various bias levels. The diode connected bias circuit is required to charge capacitive nodes in the circuit.

The practicality of this approach is highlighted even more when one considers the issue of threshold voltage shift in the chapters ahead.

8.2.2   Threshold Voltage Shift Compensation

Disordered semiconductor based TFTs experience a bias dependent threshold voltage shift in time. This shift in threshold voltage was studied and modeled in the earlier chapters. In some analog circuits e.g., light emitting diode (LED) driver circuit for displays, the threshold voltage shift in the TFT is a source of error and needs to be compensated for. The problem is that the threshold voltage of the TFT is an intrinsic parameter. Since we only have access to the drain current, gate-source voltage and drain-source voltage, the threshold voltage of the TFT for all practical purposes, is unknown and needs to be estimated during circuit operation.

The diode configuration of the TFT provides a means to estimate the threshold voltage of the transistor. If the TFT is sourced a specific current,

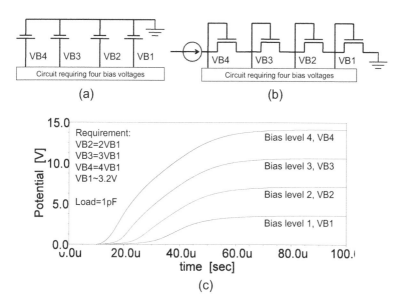

FIGURE 8.3

(a) Circuits requiring many bias lines must either be provided with independent supplies, (b) or the bias provision can be made more efficiently by the use of a chain of diode connected TFTs of specific aspect ratios driven by one independent current source. (c) Simulation of the dynamics of bias generation. At time zero, the current source is set to the appropriate value and the dynamics is the response time of the diode chain to charge the bias lines.

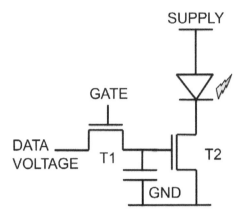

FIGURE 8.4
A two transistor LED driver circuit using TFTs.

the gate terminal of the TFT acquires a unique potential in accordance to Equation 8.3. If the threshold voltage is a function of time, then so is the voltage across the diode TFT according to Equation 8.3. In order to estimate the degree of threshold voltage shift, we only need to study the variation in V_{ds} with time. Since this is an extrinsic parameter that we have access to, it makes the estimation of threshold voltage more feasible.

This concept is best understood by the following example. Consider the driver circuit used for LED displays as shown in Figure 8.4. The circuit drives a LED with TFT T2 behaving like a current source. First, the pixel access switch (TFT T1) is closed and data to be displayed is stored on the storage capacitor, after which T1 is open. Now the data which is present at the gate of TFT T2 programs T2 to sink a specific amount of current thereby controlling the brightness of the LED. Consider the influence of the threshold voltage shift on TFT T2. Since T2 gets a random sequence of data on its gate, the threshold voltage shift in T2 is unknown at any given point in time. Thus, the increasing threshold voltage in T2, causes the LED brightness to change in time for the same applied data voltage.

The problem can be resolved by using the diode configuration of the TFT. We replace the circuit of Figure 8.4a with the circuit of Figure 8.5. Instead of programming the gate of T2 with a data voltage, we program with a data current. The circuit of Figure 8.5 includes a new TFT switch T3 which shares the same gate line as T1. During programming both T1 and T3 are closed, resulting in T2 having a diode configuration. The data current now runs through the diode connected T2. Since there is a threshold voltage shift in T2, the gate of T2 sets itself to the compensated voltage required according to Equation

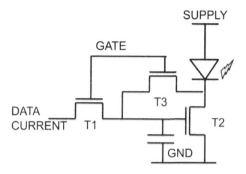

FIGURE 8.5

A modified LED driver circuit using a diode connected TFT for threshold voltage shift compensation.

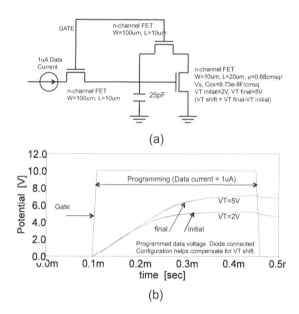

FIGURE 8.6

(a) Application of the diode TFT as a compensation circuit for the threshold voltage shift in a driving TFT (here the TFT with W=10um and L=20um). These circuits find application in LED display architectures as will be discussed in the chapters ahead. (b) Simulation of the dynamics of circuit operation.

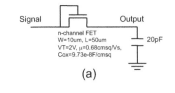

(a)

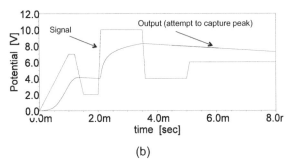

(b)

FIGURE 8.7
(a) Use of the diode-capacitor series circuit as a peak detect. (b) Simulations of the dynamics of the peak detect operation of the diode-capacitor series circuit.

8.3 to ensure that T2 drives the data current. Once T1 and T3 are switched off, the gate of T2 has the correct compensated data voltage required to ensure that the same data current is sourced from the LED. Figure 8.6 shows an illustration of this programming operation, where the gate of T3 is charged to the appropriate voltage (compensating for threshold voltage shift) in order to drive the data current. This is one illustration where the diode configuration of the TFT can be used to ascertain and compensate for threshold voltage shift.

8.2.3   Peak Detect Circuit

Many applications of large area electronic systems are related to the area of sensing where the sensor interfaces with the TFT as a switch. Signals from the sensor are readout by external electronics via the TFT switch. The signal sensed maybe time varying and sometimes it may be desired that only the peak value of the signal be read.

The diode connected TFT is a useful tool in picking the peak of the incoming signal simply due to the nature of its operation as discussed earlier. Let us consider a voltage signal being submitted to a circuit having a diode connected TFT in series with a capacitor. The signal appear at the drain (and hence gate) electrode of the TFT while the source of the TFT is connected to the capacitor. The other plate of the capacitor is considered ground. Consider two subsequent time points, t_1 and $t_2 > t_1$ on an incoming voltage signal where the value of the signal is $v(t_1) > V_T$ and $v(t_2) > V_T$. Here V_T is the threshold voltage of the TFT and is assumed constant at the subsequent time points t_1 and t_2. We assume that at t_1 the capacitor has zero charge on it. Thus, at t_1, the gate-source voltage of the TFT is $v(t_1)$ and the TFT remains on (closed) till the capacitor is charged to a voltage $v(t_1) - V_T$. After this, the gate-source voltage of the TFT is $< V_T$ and the rate of charging of the capacitor slows down to almost negligible values. Now at time t_2, the signal submits a voltage $v(t_2)$ at the drain of the diode connected TFT. The gate-source voltage of the TFT is therefore $v(t_2) - v(t_1) + V_T$ since the capacitor (and hence the source of the TFT) has a voltage $v(t_1) - V_T$ on it. If $v(t_2) > v(t_1)$, the capacitor continues to charge up to $v(t_2)$. Else if $v(t_2) \leq v(t_1)$, the voltage of the capacitor does not change and remains at $v(t_1)$. In essence, the circuit updates the voltage on the capacitor to match the highest value of the incoming signal and therefore performing a task of peak detection.

Figure 8.7 shows AIM-SPICE simulations of the peak detect operation by a diode connected TFT on an incoming time varying signal. The diode TFT attempts to capture the value of the highest possible value of the waveform. In this context it must be noted that the diode connected TFT can perform this operation without an external power supply, while it draws power from the signal itself. Thus it becomes a useful tool in large area electronic system design with disordered semiconductor based TFTs.

8.3 Conclusion

This chapter introduced the diode connected transistor. This circuit where the gate of the transistor is connected to the drain plays a useful role in biasing. With respect to the threshold voltage shift, we showed by example how this circuit could be used to actively estimate the threshold voltage and compensate for it. Thus, this circuit is a valuable entity for large area electronic systems with non-crystalline semiconductors.

9

Basic Circuits

CONTENTS

Over the years there have been several attempts at building building blocks for TFT circuits and to characterize their performance [130]-[144]. In this chapter we look at several commonly used circuits with field effect transistors. These circuits are not particularly designed for non-crystalline semiconductors but are essential building blocks for any application related to large-area electronics. However, keeping in mind the limitations of TFTs - namely threshold voltage shift, poor mobility, and lack of complementary devices, we restrict ourselves to only certain circuit architectures. This chapter justifies itself due to the importance of these elementary building blocks and their use in synthesising any system. The chapter introduces these circuits to the unfamiliar reader.

9.1   Analog and Digital Circuits

We classify the circuits into two categrories - *Analog Circuits* and *Digital Circuits*. From the point of view of the device physics there is no difference between an analog circuit and a digital circuit. However, the difference is in way signals are treated. In the case of analog circuits, the "signal space" is analog and all possible values of the signal are possible within the established boundaries. In the case of digital circuits, the signal space is discrete, and the number of discretizations or states may vary. In the case of two states, the signal may be *high* or *low*. High implies that the signal has a voltage equal to the supply voltage, and low implies that the signal voltage is ground. Since there are only two states, the digital signal is much more robust to noise, and parasitics such as fabrication process variations as compared to analog circuits. In the case of digital circuits the TFT mainly operates as a switch and is either on or off while in the case of analog circuits the TFT needs to operate in a more continuous mode across various bias conditions.

In our discussion below, we introduce single stage voltage amplifiers, current mirrors, differential amplifiers and the ring oscillator under the label of analog circuits. We then introduce the digital inverter, concepts between logic gate design, and the shift register under the label of digital circuits. We limit ourselves to just these circuits as they find applications in large-area electronic systems.

9.2 Current Mirrors

The *current mirror circuit* is shown in Figure 9.1. The primary function of the current mirror is to replicate the input current for purposes of biasing in a larger circuit.

Consider the n-type TFT current mirror shown in Figure 9.1. Both TFTs T1 and T2, have the same aspect ratio and are biased in saturation. The current through TFT T1 is I_i, and the gate voltage of T1 and T2 is $(I_i/\beta_1)^{1/2} + V_{T1}$, where V_{T1} is the threshold voltage of TFT T1. If T1 and T2 have the same threshold voltage i.e., $V_{T1} = V_{T2}$, since $\beta_1 = \beta_2$, the output current through T2, $I_o = I_i$. In general, the output to input current has a ratio of β_2/β_1.

9.3 Voltage Amplifiers

The primary function of *voltage amplifier*, is to amplify an incoming voltage signal. In large-area electronics, signals from a sensor can be amplified by voltage amplifiers in a pixel circuit before they are read out.

All TFT voltage amplifiers can be seen to have two parts — the *load* and the *driver*. The driver TFT converts the incoming voltage signal to a current signal, which then develops the output voltage across the load. Therefore, the

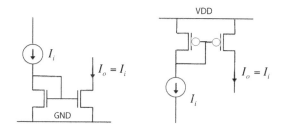

FIGURE 9.1
Current mirror.

gain of the amplifier is proportional to the voltage to current conversion gain of the driver (its transconductance), and the current to voltage conversion of the load (its impedance).

We discuss three primary architectures of single stage amplifiers — the *common source, common drain, common gate.* The term "common" is used to imply ac ground. In all the amplifier circuit, we study the low frequency gain and the output impedance.

9.3.1 Common Source Amplifier

The common source amplifier consists of a driver TFT whose source terminal is connected to dc. All TFTs are biased in saturation. Common source amplifiers are primarily used in circuits to provide signal gain.

Figure 9.2 shows the common source amplifiers with three types of loads — resistor, non-complementary (load TFT and driver TFT are both n-type or both p-type) and complementary (load and driver TFT are complementary, i.e., n-type driver and p-type load or vice-versa). Using small signal analysis, the impedance of the common source amplifiers with the n-TFT driver are $R_L||r_{od}$, $1/g_{mL}||r_{od}$, and $r_{od}||r_{ol}$, for the resistive, non-complementary and complementary loads, respectively. Recollect that r_o is the output impedance of the TFT under saturation and this resistance is due to channel length modulation. Therefore, is it typically very high. Here the subscripts d and l are used for the driver and the load TFT, respectively. The small signal gain (v_o/v_i) of the three stages are $-g_{md}(R_L||r_{od})$, $-g_{md}(1/g_{mL}||r_{od})$, and $-g_{md}(r_{od}||r_{ol})$, respectively. It must be noted that the common source amplifier is inverting, i.e. the output has a negative sign on it (phase shifted from the input by 180 deg).

In the case of using resistor loads, R_L must be made very large in order to increase the gain. This is at the cost of a large voltage drop across it which in turn allows lesser room for the driver TFT to operate in saturation and thereby implying a lower operation range for the amplifier. In the case of a non-complementary load, the operation range can be increased, but

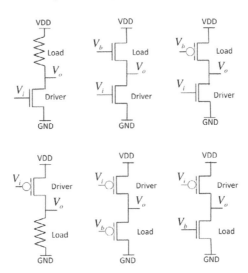

FIGURE 9.2
Common source amplifiers with resistive, non-complementary TFT and complementary TFT load.

the gain is limited because of low output impedance of the circuit $\approx 1/g_{ml}$. However, the gain $-g_{md}/g_{ml} = -((W_d/L_d)(L_l/W_l))^{1/2}$ is easily controlled by designing the aspect ratio of the load and driver appropriately. In some TFT technologies such as those based on amorphous hydrogenated silicon (a-Si:H), this architecture may be the only choice available due to the absence of good complementary devices. In the case of a complementary load, the output impedance is maximum and the gain is very high. This is the true strength of complementary MOSFET technology. Finally, due to the high gain of common source amplifiers, the Miller effect on the gate-drain capacitances becomes very prominent and limits the frequency response of the circuit.

From the point of view of threshold voltage shift, the currents through all these circuits decrease with time. However, we show in later chapters that the use of non-complementary architectures are immune to the threshold voltage shift.

9.3.2 Common Drain Amplifier

The architecture of the common drain amplifier is shown in Figure 9.3. All TFTs are biased in saturation. Common drain amplifiers are also called source followers and do not give voltage gain (i.e their voltage gain ≤ 1). However, due to their low output impedance they are used as output stages of circuits.

Once again the architecture consists of a load and a driver. Note the differ-

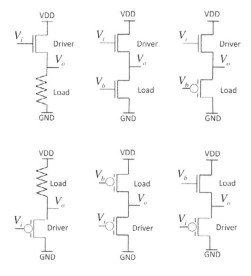

FIGURE 9.3
Common drain amplifier with resistive, non-complementary TFT and complementary TFT load.

ence from the common source amplifier in terms of the output impedance of the driver. In the case of the common source amplfier, the output impedance of the driver was very high and equal to r_{od}. In the case of the common drain amplfier the source of the driver TFT is the output node and is fluctuating. This in turn results in a gate-source voltage fluctuation and a current through the driver TFT. Therefore, the output impedance of the driver is low and equal to $1/g_{md}$. Therefore, irrespective of the load and driver configuration used, the gain is always ≤ 1. The output impedance of the common source amplifier for a resistor, non-complmenetary and complementary load is $R_L||1/g_{md}$, $r_{oL}||1/g_{md}$, and $1/g_{md}||1/g_{ml}$, respectively. The gain of the common drain amplifier is $g_{md}/(g_{md} + 1/R_L)$, $g_{md}/(g_{md} + 1/r_{ol})$, and $g_{md}/(g_{md} + g_{ml})$ for the resistive, non-complementary and complementary load, respectively. Note that the common drain amplifier does not invert the signal at the output like the common source amplifier.

Also, the Miller effect does not amplifiy the capacitances in the common drain amplifier as much as in the case of the common source amplifier due to the low gain. Therefore common drain amplifiers have good frequency response.

Once again the threshold voltage shift lowers the current through the circuit with time. However, common drain amplifiers with non-complementary load can be shown to be resistant to threshold voltage shift.

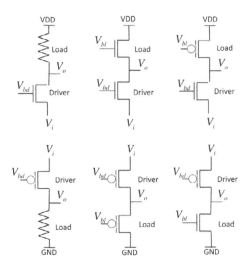

FIGURE 9.4
Common gate amplifiers with resistive, non-complementary TFT and complementary TFT load.

9.3.3   Common Gate Amplifier

The architecture of the common gate amplifier is shown in Figure 9.4. All TFTs are biased in saturation. Note that the input terminal of the common gate amplifier sees a low impedance even at low frequencies. This is unlike the common source and common drain amplifiers where the incoming signal sees a high impedance at low frequencies due to the gate capacitance. Moreover, the input impedance of the common gate amplifier can be modulated by changing the aspect ratio of the driver TFT. Since this input impedance is mainly resistive, it almost does not vary with the frequency of the voltage input. Therefore, common gate amplifiers find applications where impedance matching is required for maximum power transfer.

The low frequency input impedance of the common gate amplifier is $1/g_{md}$. The output impedance of the common gate amplifer is $R_L||r_{od}$, $1/g_{mL}||r_{od}$, and $r_{od}||r_{ol}$. The gain of the common gate amplifier is $g_{md}(R_L||r_{od})$, $g_{md}(1/g_{mL}||r_{od})$, and $g_{md}(r_{od}||r_{ol})$, for the resistive, non-complementary and complementary load, respectively.

Thus, the common gate amplifier has a high gain (like the common source) and high output impedance (like the common source), but is non-inverting. The primary difference from the common source amplifier is that the input impedance of the common gate amplifier is low.

9.4 Digital Inverter

The digital inverter is a circuit that operates in two states. The input to the circuit has two logic levels — high and low. For example, the high logic level may be represented by the input voltage having a value equal to the power supply, V_{DD} and the low logic level may be represented as the voltage being at the ground level. The function of the inverter is to "flip" or invert the input, i.e., if the input is high the output must be low and vice-versa. In the previous section on single stage amplifiers, we saw that the common source amplifier flips or inverts the input voltage. Therefore, the common source amplifier can be used as a digital inverter.

However, the common source ampifier with a resistive and non-complementary loads have a problem. Consider the driver TFT be n-type and the load to be resistive or non-complementary. When the input is low i.e., $V_i = 0$, the driver TFT is switched off, and the load lets the output equal the supply voltage V_{DD}. When the input is high i.e., $V_i = V_{DD}$, the driver TFT is on has a current through it, but so does the load. Therefore, the output ever really goes down to $V_o = 0$. Thus, while there is an inversion of input and output, the inversion is not *rail to rail* i.e., not complete. Rail-to-rail swing is very important if we have to *cascade* a long sequence of circuits operating with the same logic levels.

This problem is avoided by using a digital inverter with an architecture shown in Figure 9.5. Here the p-type TFT and n-type TFT receive the input at their gates simultaneously. As an amplifier, we may think of the n-type and p-type TFTs being a driver and the load to each other. Firstly, the gain and output impedance of this architecture is very high. Secondly, looking at the large signal operation, when $V_i = V_{DD}$, the n-type TFT is on and the p-type TFT is off, and the output is completely pulled to ground, i.e., $V_o = 0$. When $V_i = 0$, the n-type TFT is off and the p-type TFT is on and the output is completely pulled up to high $V_o = V_{DD}$, thereby achieving rail-to-rail operation.

The digital inverter may be considered to be a fundamental building block of digital circuits. Generally, it can be seen to be composed of a *pull-up* and *pull-down* unit. The pull-up unit, when "activated", pull-up the output node to high and in the case of the inverter is the p-type TFT. The pull-down unit when activated pulls down the output to low and in the case of the inverter is the n-type TFT. An important feature of the digital inverter based on complementary TFTs is that there is very little *static power consumption* when the inputs are fixed at high or low since there is no direct path between V_{DD} and ground as one of the TFTs is off. The only power consumption is due to leakage current. However, when the input changes, both TFTs are on for a short while as the input and output is in transition, and there exists a *short circuit power consumption* component during this flip. Another source

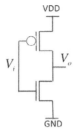

FIGURE 9.5
Digital inverter with a complementary TFT process. Any common source amplifier architecture is in general an inverter circuit.

of power consumption is the *dynamic power consumption*, which is the power needed to charge the effective capacitance, C_{eff} at the output node. The output capacitance, C_{eff}, comprises of the self capacitance of the inverter, and load capacitance of the subsequent stage. The self capacitance of the inverter consists of the Miller amplified overlap capacitances seen at the output node, and the diffusion capacitances (if any). The dynamic power consumption is the power required to charge C_{eff} to V_{DD} and is equal to $C_{eff}V_{DD}^2/2$.

An important performance parameter of the inverter is the time delay for a low-high or high-low transition at the input to be transmitted to the output. Consider a situation where the input and output conditions of the inverter are low (ground) and high (V_{DD}), respectively. We now analyze the delay in the inverter due to a low to high transition at the input. Just after the input has flipped, the p-type TFT is shut off, and the n-type TFT is on and begins to discharge the capacitor at the output. Hence we only consider the series circuit of the n-type TFT and capacitor during this analysis. First, the drain of the n-type TFT and the gate of the n-type TFT are at V_{DD} and the TFT is in saturation and operates like a constant current source independent of the drain voltage. The current through the TFT is $\beta(V_{DD} - V_T)^2$ and this current depletes the charge on the capacitor thereby lowering the output voltage. The TFT remains in saturation till the voltage on the output capacitor goes down to $V_{DD} - V_T$ from V_{DD}. The time taken for this drop in voltage is

$$t_{sat} = \frac{C_{eff}V_T}{\beta(V_{DD} - V_T)^2} \tag{9.1}$$

Once the voltage on the output node goes below $V_{DD} - V_T$, the TFT moves out of saturation and into linear mode of operation where it behaves like a resistor. The TFT continues to drain the charge out of the output capacitor lowering the output voltage from $V_{DD} - V_T$ to ground. The time taken for this drain out of the charge to occur is approximately 2.2 times the RC time constant of the capacitor-TFT series circuit. Approximating and using the

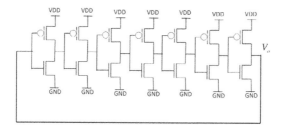

FIGURE 9.6
Ring oscillator.

small signal resistance model

$$t_{lin} \approx \frac{2.2 C_{eff}}{2\beta(V_{DD} - V_T)} \qquad (9.2)$$

The total delay of the inverter, t_D, is thus given by

$$t_D = t_{lin} + t_{sat} \qquad (9.3)$$

In order to achieve similar time constants for pull-up and pull-down operation, the equivalent resistances of the n-type TFT and p-type TFT must be made the same by sizing their aspect ratios in accordance to their mobilities. In the case of a semiconductor process which does not allow a complementary device, a diode connected TFT can be used as a load.

9.5 Ring Oscillators

The next circuit we discuss is the ring oscillator which may be classified as both an analog as well as digital circuit. It falls under a class of circuits called *oscillators* which are used to provide a periodic output waveform for applications as the *clocks* of the electronic system. Clocks are used to synchronise several events during the operation of a system. The general architecture of a ring oscillator is shown in Figure 9.6 where the oscillator consists of an odd number (≥ 3) of digital inverters connected to one another in a complete loop. Note that other than the power supply and ground terminals, there are no special input and output terminals of the ring oscillator. The ring oscillator picks up electronic noise which is then amplified by each stage around the loop till we obtain a sustained voltage oscillation from V_{DD} to ground at every node of the oscillator.

In order for a ring oscillator to oscillate, two conditions must be satisfied.

First, the total gain of all the stages after one complete loop must be ≥ 1. Second, the phase shift of the signal around the loop must be 180deg (since the loop with an odd number of inverters has a negative feedback). Suppose we have a ring oscillator of $2n + 1$ stages with each stage having a time constant $R_{eff}C_{eff}$, to meet the above two criteria, the oscillation frequency of the ring oscillator, ω_{osc}, is defined as

$$\omega_{osc} = \frac{\tan\left(\frac{\pi}{2n+1}\right)}{2\pi R_{eff}C_{eff}} \tag{9.4}$$

The low frequency gain of each stage must then be A where

$$A = (1 + (2\pi\omega_{osc}R_{eff}C_{eff})^2)^{1/2} = \sec\left(\frac{\pi}{2n + 1}\right) \tag{9.5}$$

9.6 Static Random Access Memories

Modern day memory circuits can be broadly classified as *volatile* (power supply is required to maintain the data) and *non-volatile* (power supply is not required to maintain the data). The static random access memory (SRAM) is a volatile memory and an important component of modern day computer systems. It is a memory unit into which data can be read from as well as written into.

It basically consists of two cross-coupled inverters as shown in Figure 9.7. Due to the positive feedback of the two inverters, the output node of one inverter is high and the output node of the other is low, and this state is latched till it is changed. In order to read and write information from and to the memory cell, we have two access switches which help isolate the SRAM pixel in the array. The data lines or *bit lines* (BL and BLc) are used to read and write data into the SRAM cells. The terms BL and BLc are for bit line and bit line complement as we shall see the idea behind the nomenclature during the write operation discussed below. The access switches are controlled by the gate lines or *word line* (WL).

The design of the transistors is crucial for proper operation since we do not want the data in the SRAM to be corrupted or changed (or flipped) as we access the SRAM to read, while at the same time we want the data to change when we need to write onto the SRAM. We discuss one type of read/write operation and the associated design in a SRAM cell in Figure 9.7. Let the node A contain a high or V_{DD} while the node B contains a low or ground initially. The read operation in the SRAM is as follows. Both bit lines are first charged to V_{DD} and WL is pulled high to turn the switches on to access the pixel. Nothing interesting happens between node A and the bit line, since both are

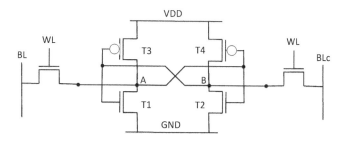

FIGURE 9.7
SRAM unit cell.

at V_{DD}. However, the bitline attempts to charge node B to V_{DD} and node B tries to discharge the bitline to ground. Since we are attempting to read the data in the SRAM, we do not want to flip the data on node B since it will flip the entire cell. Therefore we need to ensure that node B wins this "battle" and remains at ground. Thus TFTs T1 and T2 must be designed to be stronger than the access switches. This can be done by making their aspect ratios larger. Thus, T1 tried to discharge node B to ground and wins over the access switch which tries to charge node B to V_{DD}. The bit line (BL) eventually gets pulled down towards ground. External electronics containing voltage amplifiers sense the voltage swing between BL and BLc and identify the side which is logic low thereby reading the information stored in the SRAM. In order to write information into the SRAM, we must design the cell that flips to the data sent through the bitlines. Note that it is now impossible to write V_{DD} into the SRAM cell since we have designed T1 and T2 to be more stronger than the switches. Therefore, when we write into this SRAM cell, we need to write a logic low. If the bitline is at logic low or ground potential, it attempts to flip the node containing V_{DD} to ground. Since the p-type TFTs T3 and T4 try to keep the nodes at V_{DD}, we design the switches to be stronger than the p-type TFT thereby permitting a write operation. Designing SRAMs in TFT technology is possible as long as we have a complementary process available. If the process is non-complementary, it becomes highly challenging to design an SRAM cell. Even with complementary TFT processes, the differential threshold voltage shifts between the n-type TFT and p-type TFT can cause the outputs to fluctuate and permit an unwanted flipping of the SRAM cell.

9.7   Logic Gates

Logic gates are a means to implement functions of logic and are very useful circuits in digital electronic systems. It is expected that the reader have some basic knowledge on the different logic functions. Logic gates can be classified as *combinational logic gates* and *sequential logic gates*. Combinational logic implies that the logic that is evaluated is a function of the inputs at the time of evaluation alone. Sequential logic has a memory element, where the logic evaluated is a function of not only the present set of inputs but also the history. One example of a sequential logic circuit is the SRAM. In this section we look at only combinational logic design. Combinational logic circuits can be classified into *static logic* circuits and *dynamic logic* circuits. In the case of static logic, the output is evaluated at all time. In the case of dynamic logic, the output is evaluated in synchronization with an evaluate signal. We briefly look at both architectures.

9.7.1   Static Logic Gates

One of the most popular styles of designing static combinational logic gates is by the use of a pull-up and pull-down network. Depending on the inputs present there exists either a path from the output to V_{DD} or output to ground (but not both). In the complementary style of logic gate design, the pull-down network consists of n-type TFTs and the pull-up network consists of p-type TFTs. Thus, the digital inverter may be thought of as a logic gate with the pull-up and pull-down network composed of one TFT.

In general, for two inputs (say A and B), the AND logic, i.e., $A.B$ can be represented by two switches (TFTs) in series with the inputs controlling the gate of the TFTs. The OR logic, i.e., $A + B$, can be represented by two TFTs in parallel with the inputs controlling the gates of the TFTs. If a complement of an input is required we use a digital inverter. With these combinations of AND, OR, and inversion (represented by the overline) we can represent all logic.

The synthesis of a logic gate to evaluate any general logic is best learnt through examples. First, let us consider the NAND gate, i.e. $\overline{A.B}$. NAND is a universal gate from which all logic can be constructed (so is NOR). The expression under the overbar i.e., the complement, (in this case the complement of $A.B$) is the design for the pull-down network, since when this expression holds true or high, the output of the logic gate is false or low. Thus, the pull-down network consists of two n-type TFTs in series with their gates controlled by A and B. In order to design the pull-up network, we express the elements of the logic function as a function of the complement of inputs (i.e. $\overline{A}$ and $\overline{B}$).

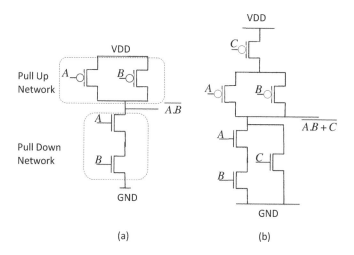

FIGURE 9.8

Static combinational logic gate with complementary process. (a) NAND gate, (b) an example for evaluating the logic $\overline{A.B + C}$.

In order to do this we use De Morgan's theorem,

$$\overline{A.B} = \overline{A} + \overline{B} \qquad (9.6)$$
$$\overline{A + B} = \overline{A}.\overline{B}$$

In the case of the NAND gate, $\overline{A.B} = \overline{A} + \overline{B}$, the pull-up network having p-type TFTs consists of two p-type TFTs in parallel, with the inputs to their gates being A and B. Thus, the NAND gate is designed as shown in Figure 9.8a. Next consider the example with three inputs A,B, and C with the logic $\overline{A.B + C}$. The first step in designing the logic gate would be to design the pull-down network which must evaluate $A.B + C$ as shown in Figure 9.8b. Next, we express the logic function in terms of the complement of the inputs i.e., $\overline{A}$, $\overline{B}$, and $\overline{C}$. Using De Morgan's theorem we see that $\overline{A.B + C} = \overline{A.B}.\overline{C} = (\overline{A} + \overline{B}).\overline{C}$. Therefore we now design the logic function $(\overline{A} + \overline{B}).\overline{C}$ with the pull-up network using p-TFTs. The logic gate is as shown in Figure 9.8b. In order to achieve symmetric performance, the TFTs can be scaled depending on whether they are in series or parallel, to match the effective performance of the inverter.

The complementary static design logic style ensures that the output is rail to rail and ensures that there is no static power consumption.

The downside to this style of logic is that to implement very large functions with n inputs, we need $2n$ TFTs. Moreover, this style of logic demands a complementary process. An alternative to this is the *Ratioed logic* architecture, which is shown in Figure 9.9 where we use a non-complementary process. If

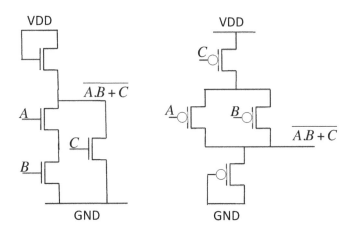

FIGURE 9.9
An example for evaluating the logic $\overline{A.B + C}$ with ratioed logic gate architecture.

the semiconductor process allows only n-type TFTs, the logic design can have one diode connected TFT as the part of the pull-up network. This TFT enures that the output is precharged to logic high. If the inputs are such that the pull-down network is activated, the pull-down network is made stronger than the diode connected TFT to ensure that output is pulled down to logic low. The advantages of this approach are that we use only $n+1$ TFTs to implement a logic with n inputs and we do not need complementary devices. However, the approach has disadvantages in that the output is no longer rail to rail since the diode connected TFT is always on, and the power consumption in the circuit is significant when the pull-down network (or the pull-up network if p-type TFTs are beings used) is turned on.

Another style of static logic gate design is the *Pass Transistor Logic* architecture, which is not discussed here.

9.7.2   Dynamic Logic

Another approach to combinational logic gate design is to evaluate the output of the logic gate only when it is required. This is accomplished by the aid of a clock. The basic architecture is shown in Figure 9.10. Consider the n-type TFT circuit, the logic gate consists of a precharge TFT and an evaluate TFT with a pull-down network in between. The circuit operates in two phases. When the precharge clock CLK1 goes high, the output node is charged to VDD. When the logic is to be evaluated, the evaluate clock CLK2 goes high. If the pull-down network is on, the output discharges to ground through the pull-down network and evaluation TFT, else the output remains high. In the

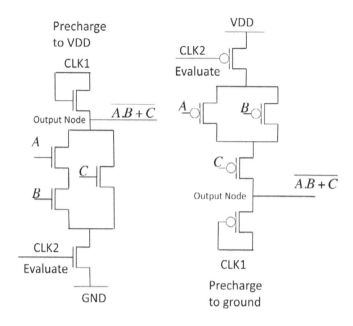

FIGURE 9.10
An example for evaluating the logic $\overline{A.B + C}$ with dynamic logic gate architecture.

case of the p-TFT architecture, the output node is precharged to ground, and depending on whether the pull-up node is turned on or not, the logic is evaluated during the evaluation cycle.

9.8   Shift Registers

Shift registers are digital circuits that "shift" or "scroll" a data signal applied at an input end, through a sequence of output nodes. They are very useful circuits from the point of view of large-area electronic systems as they can be used to drive the gate lines of the array.

We discuss a simple shift register circuit that is compatible with technologies that do not allow complementary devices. A typical shift register circuit is shown in Figure 9.11. We can recognise it as a dynamic logic gate (a dynamic inverter with a diode connected TFT load). The shift register is comprised of repeating units connected in series as shown. Each unit consists of six TFTs. The operation of the shift register is as follows. First the CLK1 signal goes high and precharges node A to V_{DD}. Then, the CLK1 signal goes low and the

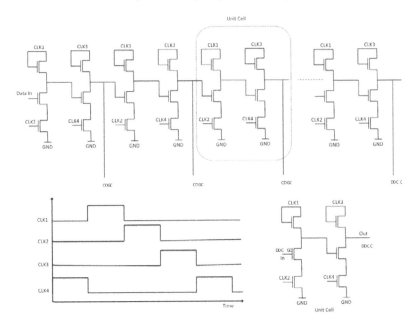

FIGURE 9.11
Shift register circuit using non-complementary TFT process. The circuit consists of unit cells in series, with each unit cell being two dynamic logic based inverters cascaded together.

CLK2 signal goes high evaluating the first circuit block. If the input is high, node A is discharged to ground. The input to this unit is the output of the previous stage, $V_{o(n-1)}$. If the input is low, node A remains at V_{DD}. Then, CLK3 goes high and precharges the output node $V_{o(n)}$. CLK3 is then made low and CLK4 is made high evaluating the second block. Thus, the output node $V_{o(n)}$, retains the value of the input. By repeating units like this the input to the shift register is scrolled through the outputs $V_{o(1)}$, $V_{o(2)}$, etc.

In the next chapter we see that this is an ideal circuit to drive the gate lines to the large-area electronic system.

9.9   Conclusion

This chapter outlined some useful circuits. The reader is encouraged to spend time thinking about their operation and identifying means to improve the design.

10

Large-Area Electronic Systems

CONTENTS

Large-area electronics implies the design of electronic systems that are spatially large. Generally this means that conventional semiconductor processes such as crystalline silicon technology become expensive and unreliable due to mismatch in device properties across the substrate. Therefore, we use electronics based on non-crystalline semiconductors that are typically fabricated at lower temperatures and over a wide variety of substrates.

In this chapter we briefly discuss the applications of TFTs based on disordered semiconductors in large-area electronic systems.

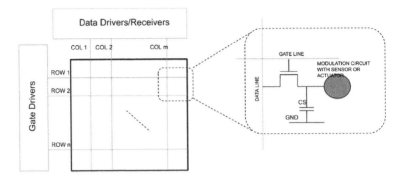

FIGURE 10.1
Typical architecture of large-area electronic systems.

10.1 Large-area Electronic Systems

So far the main applications of large-area electronics has been in displays, image sensors and tactile sensors [95], [146]-[162].

The general architecture of a large-area electronic system is shown in Figure 10.1. The system consists of an array (n by m) of pixel circuits. Each pixel may have an actuator (e.g., light emiting diode (LED) for displays) or sensor (eg. p-i-n diode for image sensor arrays). If the data has to be sent to or received from each pixel, there must be a scheme to address each pixel separately. One of the schemes of performing selection is by using a thin film transistor (TFT) as a pixel access switch. Such systems are called *active matrix* (Figure 10.1).

In active matrix large-area electronic systems, each pixel circuit has a TFT switch-capacitor circuit at the front which either directly or indirectly (through a modulation circuit) drives/receives data to/from an actuator/sensor. The gate of the access switch is driven by a gate driver circuit (eg. a shift register) housed outside the array and driving the gate lines labeled row 1 to n. The data signals are provided/received by a data driver/receiver circuit also housed outside the array through the data lines labeled col 1 to m. During operation, the jth row of the array is pulled high (assuming the TFTs are n-type) while all the other rows are kept at ground. This allows all the access TFTs along row j to remain closed while the other TFTs are open. Now, data from/to the sensor/actuator is received/sent by the data receivers/drivers via the data lines. Subsequently, the jth row is pulled to ground thereby opening the TFTs after which the $j + 1$th row is pulled high. As this operation continues, all the pixels are scanned row by row at a desired frame rate.

The most common active matrix large-area electronic systems are perhaps

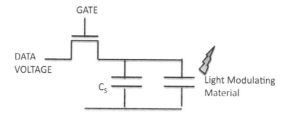

FIGURE 10.2
Pixel circuit for field-controlled actuator displays.

displays and image sensors. In the next few sections we discuss the issues involved with display and sensor larger area electronic systems.

10.2 Displays

Displays are one of the most commercially active applications of large-area electronic systems. This is even more so since modern day electronics intensively use displays as the primary human interface systems.

10.2.1 Design of Pixel Circuits for Field and Current Controlled Actuators

Active matrix displays can be classified based on the operational principle of the actuator. Actuators such as liquid crystal displays (LCDs) that are light transmission modulators, and electronic ink (e-ink) that are light reflection modulators are controlled by a field. On the other hand LEDs which are light emitters require current control. The pixel architectures for these two classifications is thus different.

10.2.1.1 Field Controlled Actuators

In the case of field or voltage controlled actuators, the pixel circuit can be as simple as a switch and capacitor circuit shown in Figure 10.2. The switch is turned on and the data line charges the capacitor to a *data voltage* that generates the required field for the actuator.

10.2.1.2 Current Controlled Actuators

In the case of current controlled actuators, the switch capacitor circuit must be followed by a current controller circuit, which in its simplest form is a single TFT operating in saturation. Consider the display pixel circuit shown

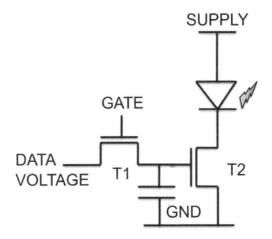

FIGURE 10.3

Pixel circuit for current controlled actuator displays. Here we use the example of an organic LED display. (From Sambandan, S.; Nathan, A.; Single-Technology-Based Statistical Calibration for High-Performance Active-Matrix Organic LED Displays, Journal of Display Technology, Volume: 3, Issue: 3 Digital Object Identifier: 10.1109/JDT.2007.900914 Publication Year: 2007, Page(s): 284–294. With permission.)

in Figure 10.3. The pixel circuit drives a LED with TFT T2 behaving like a current source. First, the pixel access switch (TFT T1) is closed and data to be displayed is stored on the storage capacitor, after which T1 is open. Now the data that is present at the gate of TFT T2 programs T2 to sink a specific amount of current thereby controlling the brightness of the LED.

Here there is an implicit assumption that the transfer characteristics of the current driver TFT is well defined. Such an assumption is not correct due to the threshold voltage shift. If a data voltage V_{DATA} is written onto the storage capacitor, the current through the LED is $\beta_{T2}(V_{DATA} - V_T^{T2})^2$, where V_T^{T2} is the threshold voltage of TFT T2. Due to threshold voltage shift in TFT T2, V_T^{T2} is unknown and hence the current through the LED cannot be foreseen when the data voltage is applied.

A simple technique of compensating for this deficiency is to program the circuit with the desired current itself, thereby letting the gate of the current driver TFT set itself to the appropriate value independent of the threshold voltage shift. Consider the circuit shown in Figure 10.4. It includes a new TFT switch T3 which shares the same gate line as T1. Instead of programming the gate of T2 with a data voltage, we program the circuit of Figure 5 with a data current, I_{DATA}. During programming both T1 and T3 are closed, and

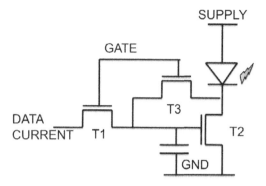

FIGURE 10.4
Threshold voltage shift compensated pixel circuit. (From Sambandan, S.; Nathan, A.; Single-Technology-Based Statistical Calibration for High-Performance Active-Matrix Organic LED Displays, Journal of Display Technology, Volume: 3, Issue: 3 Digital Object Identifier: 10.1109/JDT.2007.900914 ?Publication Year: 2007, Page(s): 284–294. With permission.)

the data current now runs through the diode connected T2. Since there is a threshold voltage shift in T2, the gate of T2 sets itself to the compensated voltage required to ensure that T2 drives the data current. In other words the gate of T2 becomes $(I_{DATA}/\beta_{T2})^{1/2} + V_T^{T2}$. Once T1 and T3 are switched off, the gate of T2 has the correct compensated data voltage required to ensure that the same data current is sourced from the light emitting diode.

In the next chapter we look into compensation circuits and strategies in more detail.

10.2.2 Design Constraints

In the design of any display system the following parameters are fixed. The frame rate or refresh rate of the display cannot be less than a certain rate F_r as the human eye is sensitive to rates below 50 Hz. The minimum feature size, L, dielectric capacitance, and overlap capacitance of the TFTs is a process parameter and can be considered fixed. Similarly the maximum gate voltage, the initial threshold voltage of the TFT, the supply voltage, and the maximum and minimum programmable data voltages are again process parameters and are fixed. Device properties of the TFT such as the field effect mobility, and leakage current dependence on the field are also considered fixed.

The circuit design parameters that are flexible are the size of the display array, the pixel size, aspect ratio of the TFTs, the size of the storage capacitor and the topology of the circuit along with the programming scheme. Considering the basic circuit architectures and programming schemes discussed above, we look at the influence of the aspect ratio of the TFTs and the size of the storage capacitor on the display performance and the performance parameters discussed below.

10.2.2.1   Programming Time

The programming time, t_p, of a single pixel is the time taken to write 90% of the data into the pixel storage capacitor. It is directly related to maximum number of rows, n, a display can have since $n = (t_p F_r)^{-1}$.

First, we consider the programming time of the pixel circuit for displays having voltage programming i.e., voltage is written onto the storage capacitor via the access switch. As the aspect ratio of the pixel access TFT increases and as the capacitance of the storage capacitor decreases, the programming time decreases. The time for programming is accurately defined by the solution to the equation

$$C_s \frac{dv_C(t)}{dt} = \frac{\mu}{2} \frac{W_s}{L_s} C_{ox}(V_{DATA} - v_C(t))(\vartheta - v_C(t)). \tag{10.1}$$

The solution of (10.1) yields

$$v_C(t) = \frac{V_{DATA}\vartheta \left(1 - e^{-\gamma t}\right)}{\vartheta - V_{DATA}e^{-\gamma t}} \tag{10.2}$$

where $\vartheta = 2(V_{SEL} - V_T) - V_{DATA}$, and $\gamma = (\vartheta - V_{DATA})\frac{W_s}{L_s}\frac{\mu C_{ox}}{C_s}$ where one can identify γ^{-1} to be the small signal time constant of the RC circuit formed by the select TFT and the storage capacitor. V_{SEL} refers to the gate voltage of the pixel select switch when the pixel is accessed.

In the case of the current programmed circuit discussed for LED displays with compensation for threshold voltage, the input stage typically consists of a diode connected TFT that helps convert the data current into the appropriate voltage to be stored on the capacitor. During programming, this input diode connected TFT acts as a nonlinear resistor in parallel with C_s and the charging of the capacitor is governed by the equation

$$I_{DATA} = I_{LEAK} + (C_s + C_{LINE})\frac{dv_C(t)}{dt} + \frac{\mu C_{ox}}{2}\frac{W_d}{L_d}(v_C(t) - V_T)^2, v_C > V_T. \tag{10.3}$$

The transients of v_C is given by

$$v_C(t) = \left[\left(\frac{I_{DATA} - I_{LEAK}}{\frac{\mu C_{ox}}{2}\frac{W_d}{L_d}}\right)^{\frac{1}{2}} + V_T\right]\frac{1 - e^{-\omega t}}{1 + e^{-\omega t}} \tag{10.4}$$

where $\omega = \dfrac{2\left(\frac{\mu C_{ox}}{2}\frac{W_d}{L_d}(I_{DATA} - I_{LEAK})\right)^{\frac{1}{2}}}{C_s + C_{LINE}}$. The time required for programming the pixel is dependent on the value of the data current and is large for small

I_{DATA} and the presence of I_{LEAK} makes this a serious issue in the case of large diagonal displays.

As the display diagonal increases, the voltage programmed pixel circuit is the more appropriate choice unless a suitable current driving circuit can be implemented.

10.2.2.2 Power Consumption

Field controlled actuator pixel circuits do not consume significant power when on. All the power consumption in these circuits is during the dynamics of programming which results in the charging of the capacitor. If the maximum data voltage is $V_{DATAmax}$ the maximum power consumption per programming is $\frac{C_s V_{DATAmax}^2}{2}$.

On the other hand, current controlled actuators such as LEDs which are emissive have not only dynamic power consumption but also have static power consumption during the operation of the display. TFT circuits typically operate with about 10V supplies and consume about 1uA of current resulting in uW power consumption per TFT. However, with millions of pixels operating simultaneously in displays the power static power consumption can be significant.

10.2.2.3 Leakage

Reducing the programming time can help increase the display size. However, there is a limit to deciding the number of rows in the display and this is due to the leakage current in the pixel circuit. After programming the pixel, we have a data voltage $0.9V_{DATA}$ written on the storage capacitor.

Let the leakage through the pixel select TFT be I_{LEAK}. After one frame cycle data is lost from the capacitor due to leakage. If we impose a condition that the data lost due to leakage must not be greater than a factor of 10% of the programmed voltage, we have a constraint

$$\frac{I_{LEAK}}{F_r C_S} \leq 0.09 V_{DATA} \tag{10.5}$$

The leakage current is a function of the applied data voltage, the channel width and channel length of the pixel select TFT. According to the Poole-Frenkel relationship, the leakage current is approximately given by $I_{LEAK} = I_{L0} W_s e^{b(V_{DATA}/L_s)^{1/2}}$, where I_{L0} is the leakage per unit length at small potential drops, and b is a constant coefficient.

10.3   Sensors

Another major application of large-area electronic systems is in the development of sensors to sense physical parameters across large spatial areas. We consider the case of any general sensors to discuss the design issues related to pixel circuits for sensing.

10.3.1   Design of Pixel Circuits for Sensors

The main task of the sensor is to convert physical information (eg. pressure, temperature, light intensity etc.) to electron charge. This charge is placed on a storage capacitor, which is accessed by external circuits via a pixel access TFT switch. A typical pixel circuit for sensing applications is shown in Figure 10.5. The operation of the circuit is as follows. During the sensing cycle, the sensor dumps charge on the storage capacitor for the time equal to the frame rate of the readout cycle. During readout, the pixel select transistor is turned on, and the charge stored on the capacitor is shared with the data line. After charge sharing, the pixel select transistor is closed, and the data is read out from the data line by the external readout circuit. The charge transferred to the data line is $\frac{C_{LINE}}{C_S+C_{LINE}}Q_{sig}$, where Q_{sig} is the charge stored on the pixel storage capacitor before readout, C_{LINE} is the data line capacitance, and C_S is the pixel storage capacitance. The data line capacitance is mainly comprised of the parallel combination of the overlap gate-source/drain and diffusion capacitances of all the pixel select transistors connected to the column. For large arrays, this capacitance is much bigger than the pixel storage capacitor and the charge transferred to the data line is almost Q_{sig}. Since the data line capacitance is usually very big, its voltage swing due to the sensor charge is very small (Q_{sig}/C_{LINE}). Hence, the readout circuit must typically consists of a charge amplifier. After the pixel select TFT is turned off, the remaining charge on the pixel storage capacitor is removed via a reset signal provided to the reset TFT.

A modification to the above pixel circuit is the use of a source follower amplifier as the output stage as shown in Figure 10.6. The use of the source follower is beneficial in some ways as it helps increase the signal on the data line. The voltage on the storage capacitor just before readout is Q_{sig}/C_S, which is a small signal input to the gate of the source follower driver TFT. During read out the data line is typically precharged to at least one threshold voltage below the source follower drive TFT to ensure proper operation. During readout, the source follower will attempt to charge the data line to a differential potential of $\approx AQ_{sig}/C_S$, where $A < 1$ is the small signal gain of the source follower amplifier. The term "differential potential" indicates the impact of the signal on the data line i.e., the difference in the data line potential when $Q_{sig} = 0$ as compared to present value. If $A \approx 1$, the charge on

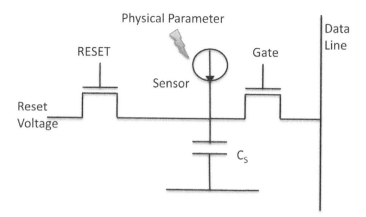

FIGURE 10.5
Pixel circuit for large-area electronic sensors.

the data line becomes, $C_{LINE}Q_{sig}/C_S$, which is much higher than the charge without the source follower output stage (which is Q_{sig}). This improves the signal to noise ratio provided the source follower itself is not too noisy. The cost for this improvement is layout area and the pixel readout time.

10.3.2 Noise

Noise considerations are extremely important when designing large-area electronic circuits for sensor applications. The accuracy of the sensor is dependent on the signal-to-noise ratio (SNR) which is the ratio of the signal power to noise power (which for example, are measured in V^2 for electronic noise). There are several sources of noise in any general sensor pixel circuit.

10.3.2.1 Noise from the Sensor

Depending on the type of sensor used, the sensor is the primary source of noise. For example, in the case of image sensors based on pin photodiodes, the photodiode is a source of dark current shot noise.

10.3.2.2 Noise from Reset

The signal integrated on a pixel is measured relative to its reset level. The thermal noise uncertainty associated with this reset level is referred to as the reset noise. The combination of the reset transistor and storage capacitor is an RC circuit. The thermal noise from the reset transistor coupled with the low pass filtering function of the RC circuit over all frequencies results in a noise element with noise power kT/C_S.

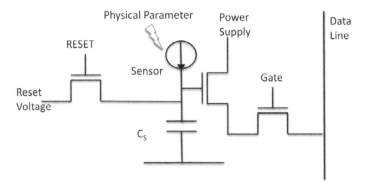

FIGURE 10.6
Pixel circuit for large-area electronic sensors with source follower output stage.

10.3.2.3   TFT Flicker and Thermal Noise

Non-crystalline semiconductor based TFTs have a large component of flicker noise due to the poor gate-insulator interface. The thermal noise and flicker noise components from the TFT (for example in the source follower amplifier) add onto the noise components.

10.3.2.4   Threshold Voltage Shift

The threshold voltage shift in the reset TFT, the source follower driver TFT and the pixel access switch are also a source of noise. The threshold voltage shift in the source follower driver TFT is particularly significant since it is spatially non-uniform, and significantly affects the data stored on the data line due to changes in the amplifier gain.

10.3.2.5   External Readout Noise

The external readout circuit is also a source of noise. Since the amplifier on the outside sample the charge on the data line, they are a source of kT/C noise. The data line is also a significant contributor of thermal noise. Moreover, the large data line capacitance converts any voltage noise into significant fluctuations in charge levels.

10.3.3   Overcoming Noise

While there are several sources of noise, some techniques can be used to improve the signal-to-noise ratio. By reducing the overlap capacitances of the pixel select TFT, (for example, by using a self-aligned fabrication process), the data line capacitance can be significantly lowered. The data line resistance can be lowered by fabrication techniques that permit high aspect ratio features. Flicker noise can be largely eliminated through rapid double sampling of the data line by the external readout circuit.

10.4 Conclusion

This chapter presented the basic concepts and design parameters involved in the design of large-area electronic systems. We saw that the threshold voltage shift in the active TFTs of the pixel circuit lead to problems in data accuracy and noise. In the next chapter we look at some of the techniques used to overcome the problems posed by the threshold voltage shift primarily for applications related to LED displays.

11

Compensation Circuits for Displays

CONTENTS

In the previous chapters we saw that the threshold voltage shift in the TFT based on non-crystalline semiconductors is a fundamental bottleneck for circuit design.

As long as the TFT is used as a switch the threshold voltage shift can be compensated for by setting the off voltage of the switch to a level that aids recovery. However, if the TFT switch is on for longer periods than it is kept off, the threshold voltage shift in the TFT will cause the transfer characteristics of the TFT to vary with time thereby affecting the circuit operation. In this chapter we discuss the use of compensation circuits to compensate for the threshold voltage shift in the TFT [164]-[176].

11.1 Compensating for Threshold Voltage Shift

Compensation or correction for threshold voltage shift in a TFT is achieved by a *compensation circuit*. The compensation circuit is a control circuit that monitors or evaluates the threshold voltage shift in the TFT of interest and offers a correction with time. Compensation circuits for threshold voltage shift compensation were mainly designed for applications related to active matrix organic LED displays. The LED is a current controlled light emitter a TFT is biased in saturation and used as a current source to surce the current to the LED. If this current sourcing TFT has a threshold voltage shift, the LED current varies. Thus compensation circuits are used to estimate this change in threshold voltage and program the current sourcing TFT to the correct value in order to operate the display properly. However, the general idea of compensation circuits can be used for any application. Certain non-crystalline semiconductors such as polycrystalline silicon do not experience significant threshold voltage shift with time. However, there occurs a spatial mismatch in threshold voltage, which too can be compensated for with compensations circuits.

There are several limitations in using compensation circuits. First, the compensation circuit must be composed of TFTs operating as switches with short effective on time. This must be so since the TFTs in the compensation circuit themselves must not be subject to the effect of their own threshold voltage shift. Second, the compensation circuit must have a small layout area by using a minimum number of TFTs and signal lines. Finally, compensation of threshold voltage shift for a TFT is practical only when there are few TFTs that require to be compensated for. In larger circuits where several TFTs work in continuous mode, compensation circuits are not a practical option.

In general, the compensation circuit needs to ensure that the TFT being corrected retains its transfer characteristics with time. In other words the TFT must source the same amount of current, irrespective of time, when a certain gate voltage is provided on its gate. Due to the threshold voltage shift in the TFT, this is possible only if the threshold voltage of the TFT is added to the gate voltage.

Compensation circuits can be broadly classified as *voltage programmable* [164]-[169] and *current programmable* [170]-[176]. In the section below we study the general idea behind the threshold voltage compensation. While there exist many different designs and addressing schemes for compensation circuits, the circuits discussed below form the basis of operation for most of them.

11.2   Voltage Programmed Compensation Circuits

In voltage programmed compensation circuits, the TFT is provided the "data" as a voltage, V_{DATA}, on its gate and is added to the evaluated threshold voltage of the TFT. There are several schemes which achieve this.

11.2.1   Capacitor-Diode TFT based Circuits

The general scheme of these circuits is described in Figure 11.1 where the TFT to be corrected is labeled TC. The gate of the TFT TC is programmed by a voltage developed across a storage capacitor. TFT TC is diode connected

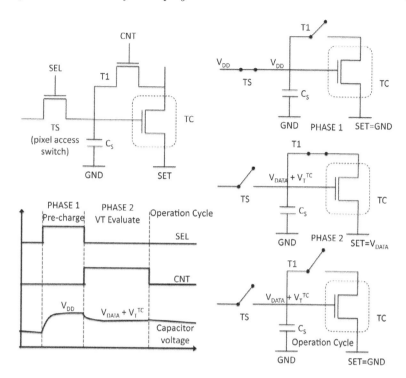

FIGURE 11.1
Capacitor-diode TFT voltage programmed pixel circuit.

with a switching TFT T1 in the path of the route between the gate of TC and the drain of TC. The source of the TFT TC is provided a signal SET.

The circuit works in three phases. The first phase is the pre-charge phase where in the storage capacitor, C_S, is pre-charged to the supply voltage, V_{DD} via a pixel select switching TFT, TS. The pixel select TFT is controlled by a gate signal labeled SEL. The SET signal is at ground. After the pre-charge, TS is opened and the switch T1 is closed by pulling up the gate control signal CNT. This phase is the threshold voltage evaluation phase. The SET signal is set at the data voltage, V_{DATA}, the TFT is desired to be programmed to. The capacitor now discharges through TC to a voltage $V_{DATA} + V_T^{TC}$, where V_T^{TC} is the threshold volatge of TFT TC at that instant in time. Thus, the threshold voltage of the TFT TC is evaluated. Finally, the copensation circuit is disconnected from TFT TC by opening the switch T1. The source of TFT TC can now be connected to ground by setting SET to ground. The TFT TC now sinks a current of βV_{DATA}^2. This current is independent of the threshold voltage shift in TFT TC (to first order).

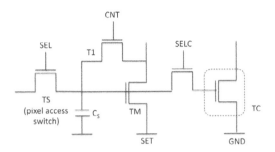

FIGURE 11.2
Mirror based voltage programming.

11.2.2   Mirror TFT based Circuits

The capacitor-diode based compensation circuits operates in two modes — programming or evaluation mode where the circuit evaluates the threshold voltage of the TFT of interest, and the operation mode where the compensation circuit is disconnected from the TFT and circuit runs as normal. This involves a break in the operation of the circuit and may not be compatible with all circuits.

In order to run the circuit in continuous mode, a TFT very similar to the TFT TC is placed close to TC as shown in Figure 11.2. This TFT (labeled TM) is called the twin of TFT TC. Both TC and TM receive the same bias conditions. With time, TFT TC experiences a threshold voltage shift. The threshold voltage of the twin TFT TM is evaluated periodically and TFT TC is disconnected during the threshold voltage evaluation via a switch controlled by signal SELC. After the threshold voltage evaluation a compensation based on this evaluation is provided to TFT TC. In this manner the operation of TC is not interrupted. The estimation of the threshold voltage of TM is performed by any suitable method (for example, using the diode based voltage programming discussed earlier).

However, this approach makes an implicit assumption that both TFT TC and its twin TFT TM experience the same threshold voltage shift. This is not a very bad assumption as we shall see in the next chapter.

11.3   Current Programmed Compensation Circuits

In current programmed compensation circuits, the TFT is directly provided the data as a current, I_{DATA}.

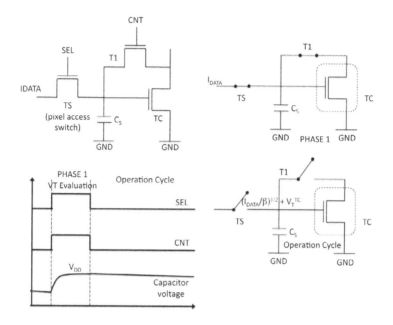

FIGURE 11.3
Capacitor-diode TFT current programmed pixel circuit.

11.3.1 Capacitor-Diode TFT based Circuits

The circuit is shown in Figure 11.3. This circuit works in two phases — the evaluation phase where the threshold voltage of TC is evaluated, and the operation phase where the TFT TC operates with the corrected data current. In the first phase the pixel select switch TS and switch T1 are closed while the data current I_{DATA} flows through the diode connected TFT TC. The capacitor builds a gate voltage $(I_{DATA}/\beta)^{1/2} + V_T^{TC}$. This gate voltage compensates for the threshold voltage of TFT TC. After evaluation, the switches TS and T1 are opened and TFT TC sources a current I_{DATA}.

11.3.2 Mirror TFT based Circuits

The capacitor-diode TFT based current programmed circuits do not operate in continuous time mode. Hence, for applications that require a continuous operation of TC, a mirror TFT based compensation circuit is used. A TFT TM identical to TFT TC is placed in a current mirror configuration as shown in Figure 11.4. The threshold voltage of the twin TFT TM is evaluated periodically using current programming exactly as described above. The TFT TC is disconnected during the threshold voltage evaluation via a switch controlled by signal SELC. After evaluation the correct threshold voltage compensated

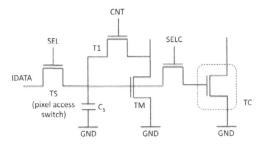

FIGURE 11.4
Mirror based current programming.

potential is provided to TFT TC thereby correcting for the threshold voltage shift.

11.4 Other Compensation Circuits for Display Applications

The above discussed compensation circuit schemes can be applied to control the current sourcing TFT in active matrix organic LED displays. Each pixel in the display array contains the compensation circuit along with the required signal lines. The downside of using compensation circuits in this fashion is the expensive layout area which results in poor display resolution. Moreover the presence of multiple TFTs in the pixel as well the presence of signal lines results in the loss of reliability.

In order to address these problems several approaches [177]-[182] have been proposed of which we discuss two techniques. The first involves the use feedback, for example, of comparator circuits for every row of the display array [177]-[180]. The second approach is more applicable to video displays and uses the statistical properties of the threshold voltage shift to compensate the pixels using software [181]. We discuss both these approaches briefly.

11.4.1 Feedback based Compensation

The general schematic of the feedback compensation circuit is shown in Figure 11.5. A current comparator is assigned each row of the display array. During evaluation, the pixel select switches controlled by the gate signal SEL are closed and the switching TFT T2 controlled by SELC is opened. This allows the current I_{TC} through TFT TC to flow to the current comparator. The comparator compares the current I_{TC} with the intented current, I_{DATA}. Any

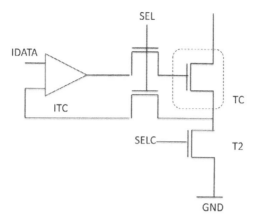

FIGURE 11.5
External programming via feedback.

mismatch in these currents generates an output voltage that is fed back to the gate of the driver TFT TC thereby programming the gate to a threshold voltage shift adjusted value and setting the current through TFT TC to the desired value of I_{DATA}. Finally, the select switches are opened and TFT T2 is closed disconnecting TFT TC from the comparator and setting the pixel in operation mode. However, this compensation scheme needs to deal with mismatches in line resistance and programming delays due to line capacitance.

11.4.2 Statistics based Software compensation

For video displays the TFTs in each pixel receive a variety of data voltages over a short period of time. Dividing the time interval into units of length τ, the cumulative effect of the sequence of data voltages on the threshold voltage of the LED driver TFT at time $n\tau$ is given by

$$V_{T,n} = V_{T(0)} + \varphi f(\tau) \sum_{i=1}^{n} (V_{DATA,i} - V_{T0})(1 - f(\tau))^{n-i} \qquad (11.1)$$

where $f(\tau)$ is a function determining the threshold voltage shift in the drive TFT in the small time interval τ. Let this sequence of data voltages received by the drive TFT, have a certain probability distribution function with mean $V_{DATAmean}$. Assuming all pixels in the display receive the data from the same distribution and mean, the mean threshold voltage shift in the TFT of any random pixel at time $n\tau$ is

$$\langle V_{Tn} \rangle = V_{T(0)} + \varphi f(\tau)(V_{DATAmean} - V_{T0}) \sum_{i=1}^{n} (1 - f(\tau))^{n-i} \qquad (11.2)$$

Thus, the mean threshold voltage shift for a TFT experiencing a variable gate bias selected from a stationary distribution for a sufficiently long time is given by the threshold voltage shift experienced by the TFT for a constant gate stress equal to $V_{DATAmean}$.

Thus, if the pixels in the display all have the initial same value, and receive data from the same distribution, after a sufficiently long period of time, all pixels in the display will have the approximately the same threshold voltage. This powerful consequence implies that one could use an external dummy TFT and evaluate the threshold voltage of that TFT for the same kind of data signals. This value of threshold voltage can then be used to blindly compensate all the pixels in the display. This method of compensation saves much layout area and reliability issues related to the design of compensation circuits in each pixel.

However, there remains a case when by chance some TFTs begin to receive data signals of a different distribution. In such a case of small deviations from the mean, the threshold voltage shift in the TFT has a self correction mechanism. Consider two TFTs, T1 and T2, with T1 having a larger initial threshold voltage compared to T2. If both TFTs are supplied the same gate voltage, we have

$$\frac{\partial V_{T1}(t)}{\partial t} < \frac{\partial V_{T2}(t)}{\partial t} \tag{11.3}$$

This implies that, with time, all TFTs will try to match the value of their threshold voltage. This property of self-mismatch-correction tends to fight the spreading of the spatial distribution of the threshold voltage due to system entropy.

11.5   Conclusion

This chapter dealt with compensation circuits to evaluate the threshold voltage of a TFT in order to compensate for the threshold voltage shift. If there is a single TFT whose threshold voltage needs to be evaluated, a compensation circuit is attached to the TFT. The compensation circuit is then addressed and operated via a certain signalling scheme, so that at the end the threshold voltage of the TFT to be controlled is evaluated. This concept of using compensation circuits is practical if there are a few TFTs that need to be controlled.

12

Self Compensation of Threshold Voltage Shift

CONTENTS

In the previous chapter we discussed the influence of the threshold voltage shift in TFTs. We saw that one of the approaches to solve the problem of threshold voltage shift is the use of compensation circuits. The role of the compensation circuit was to evaluate the threshold voltage in the TFT of interest. However, compensation circuits are a practial approach when one or very few TFTs have to be compensated for the threshold voltage shift. Typically these TFTs work in analog mode and the accuracy of the effective gate-source voltage is important — for example active matrix light emiting diode display pixel drivers. However, as the circuit gets more complex with more TFTs needing to be controlled, building compensation circuits for each TFT becomes very cumbersome.

In this chapter we take a very different view on how to tackle the threshold voltage shift. Instead of taking on a control centric approach we seek to construct circuit topologies that are inherently time invariant by their very nature [137]-[138], [141], [183]. In some sense, we see if we can *self assemble stability* in the circuit.

12.1   The Dancing House Analogy

In order to define the problem more clearly, we look at it using an example of building a house with a special kind of bricks. These special bricks change their shape with time. If we need to build a house with these bricks, the house would change shape too. However, we ask the following question. Can we lay the bricks in such a manner that they compensate for each other's change in shape thereby keeping the house's shape constant?

If we think about it, building circuits with TFTs is like building a house with shape changing bricks. The threshold voltage shift in the TFT makes the transfer characteristics of the TFT change with time and hence the TFTs are analogous to the shape changing bricks. Can we develop a theory for circuit synthesis that defines circuit topologies where in TFT circuits will have time invariant transfer characteristcis, even if the TFTs themselves have a time dependent transfer characterisitcs? The transfer characteristics or transfer function of a circuit is the ratio of the output of the circuit to the input to the circuit.

The answer to this question is in the affirmative. We approach circuit design with the TFT from a more general perspective and show that one can *self assemble the function of time invariant behavior* using TFTs.

12.2   Graphical Representation of a TFT

In this section we lay the foundations for a generic approach to circuit synthesis with TFTs. The aim is that the circuit be intrisically time invariant. This implies that the transfer function of the circuit be time invariant in spite of the TFTs having time variant transfer characteristics because of threshold voltage shift.

The first step in this direction is to build a black box representation of the TFT in terms of the inputs and outputs. We consider the TFT to always be in saturation mode of operation, and neglect the effect of channel length modulation.

12.2.1 Constant Voltage Bias

For a TFT in saturation, the current voltage relation can be expressed as

$$I_{ds} = \beta(V_{gs} - V_{T0} - \delta V_T(t))^2 \tag{12.1}$$

where V_{T0} is the threshold voltage of the TFT at time zero, and $\delta V_T(t)$ is the change in threshold voltage with time. For a constant gate bias of V_{gs},

$$\delta V_T(t) = (V_{gs} - V_{T0})f(t) \tag{12.2}$$

where $f(t)$ illustrates the time dependence of the threshold voltage shift. For example, in the case of a-Si:H based TFTs where defect state creation is the main mechanism of charge trapping, $f(t)$ is the stretched exponential function, $1 - e^{-(t/\tau)^\beta}$.

12.2.2 Constant Current Bias

Equivalently, if we bias a TFT in saturation and force a current, I_{ds}, through it while allowing the gate-source voltage free to settle to any value, the threshold voltage shift results in the gate-source voltage changing with time such that

$$V_{gs} - V_{T0} = \left(\frac{I_{ds}}{\beta}\right)^{1/2} + \delta V_T(t) \tag{12.3}$$

Under constant current bias the threshold voltage shift in the TFT is given as

$$\delta V_T(t) = \left(\frac{I_{ds}}{\beta}\right)^{1/2} \frac{f(t)}{1 - f(t)} \tag{12.4}$$

12.2.3 Equivalence of Current and Voltage Bias

While we have separately stated and defined the threshold voltage shift due to voltage and current bias, it must be understood that these are equivalent. For a TFT in saturation, the current through the TFT is caused by the presence of the gate-source bias, and a freely floating gate is set to a certain potential based on the current through the TFT. Thus, the threshold voltage shift in the TFT must turn out to be the same when considering this gate bias or when equivalently considering this current.

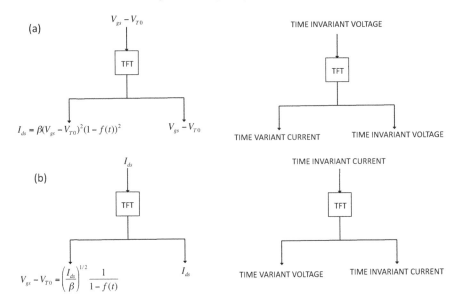

FIGURE 12.1
Modular representation of a TFT. The direction of the arrow indicates the direction of signal flow. For a given input there are two possible outputs, one corresponding to a voltage type while the other corresponding to a current.

12.2.4   Modular Representation of the TFT

Thus, from Equation 12.1 and Equation 12.2, we can rewrite the current through a TFT biased in saturation with constant gate bias as

$$I_{ds} = \beta(V_{gs} - V_{T0})^2(1 - f(t))^2 \tag{12.5}$$

From Equation 12.3 and Equation 12.4 we can write the gate-source bias across a TFT biased in saturation with a constant current as

$$V_{gs} - V_{T0} = \left(\frac{I_{ds}}{\beta}\right)^{1/2} \frac{1}{1 - f(t)} \tag{12.6}$$

We can now represent a TFT under constant gate bias and constant current bias in a modular graph node like representation as shown in Figure 12.1. In Figure 12.1a, when the input to the node is a constant voltage, the output of the node is a time varying current or another constant voltage (the input volatage itself). In Figure 12.1b, when the input to the node is a constant current, the ouptput of the node is a time varying voltage, or another constant current (the input current itself).

This node represents the TFT. It is important to note that for a constant input voltage bias, the TFT has a time invariant voltage output and a time variant current output. For a constant input current bias, the TFT has a time invariant current output and a time variant voltage output.

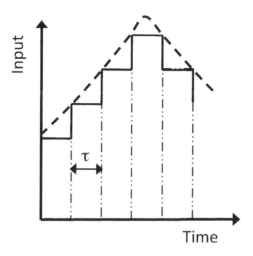

FIGURE 12.2
Time varying input conditions can be analysed by looking at small subdivisions of time of length τ.

12.2.5 Time Varying Bias

We have stated that Equation 12.5 and Equation 12.6 define the variation in the output current and output voltage under the conditions of constant voltage and current bias, respectively. However, even if the bias conditions were varying with time, the above equations can still be used to define the outputs - but with evaluation at every infinitesimal step. Figure 12.2 illustrates this idea where the time varying input bias (voltage or current) to the TFT can be broken into small steps of length τ. The time variance is stalled to evaluate the output current or voltage of TFT for that frozen moment. The evolution based on these evaluations with time would properly define the output.

12.2.5.1 Voltage Bias

This idea can be refined further using the time varying bias models for the threshold voltage shift derived in the earlier chapters. When the TFT is subjected to a sequence of different gate bias voltages the threshold voltage shift at the end of the time interval $(j-1)\tau \leq t < j\tau$ is given by

$$V_{T,j} = V_{T,j-1} + [\varphi(V_{gs-j} - V_{T0}) - (V_{T,j-1} - V_{T0})]f(\tau) \qquad (12.7)$$

where $V_{T,j}$ is the threshold voltage at the end of the time interval $j\tau$, $V_{gs,j}$ is applied gate bias during time t with $(j-1)\tau \leq t < j\tau$. In our earlier models we found that for short term varations in voltage bias, $f(\tau) = 1 - e^{-\kappa\tau}$ was a reasonably good function. Summing this recursive definition we obtain the threshold voltage at the end of the time interval $(n-1)\tau \leq t < n\tau$

$$V_{Tn} = V_{T0} + \varphi f(\tau) \sum_{j=1}^{n} (V_{gs-j} - V_{T0})(1 - f(\tau))^{n-j} \qquad (12.8)$$

12.2.5.2 Current Bias

In the case of a time varying current stress, the gate to source voltage in Equation 12.7 can be replaced by $V_{gs,j} = (I_{ds,j}/\beta)^{1/2} + V_{Tj}$. Therefore, the threshold voltage shift at the end of the time interval $n - 1\tau \leq t < n\Delta T$ is given by

$$V_{Tn} = V_{T0} + \frac{\varphi f(\tau)}{1 - \varphi f(\tau)} \sum_{j=1}^{n} \left(\frac{I_{ds,j}}{\beta}\right)^{1/2} \left(\frac{1 - f(\tau)}{1 - \varphi f(\tau)}\right)^{n-j} \qquad (12.9)$$

12.3   Simple TFT Circuits as Node Diagrams

We now see what happens when we connect up these nodes. Consider Figure 12.3 which illustrates all the possible outcomes of connecting two TFTs. We call the voltage and current outputs as V_o and I_o, respectively.

12.3.1 Voltage Input

In Figure 12.4 we have the case of TFT T1 being supplied a constant input gate voltage, $V_{gs} - V_{T0}$. The outputs of T1 as shown in Figure 1a will be

$$\begin{aligned}
V_{o1} &= V_{gs} - V_{T0} & (12.10)\\
I_{o1} &= \beta_1(V_{gs} - V_{T0})^2(1 - f(t))^2
\end{aligned}$$

These outputs of T1 can now be fed to another TFT. We feed V_{o1} as an input to TFT T2. The outputs of T2 then are

$$\begin{aligned}
V_{o2} &= V_{o1} = V_{gs} - V_{T0} & (12.11)\\
I_{o2} &= \beta_2 V_{o1}^2(1 - f(t))^2\\
&= \beta_2(V_{gs} - V_{T0})^2(1 - f(t))^2
\end{aligned}$$

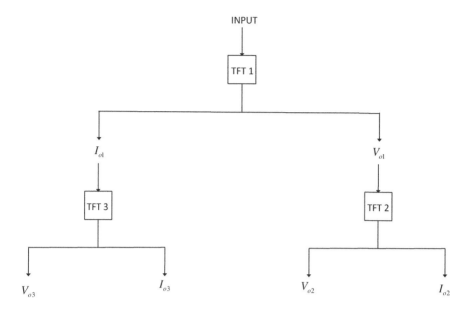

FIGURE 12.3

Circuits of two TFTs. (From Sambandan, S.; Nathan, A.; Circuit Techniques for Organic and Amorphous Semiconductor based Field Effect Transistors, IEEE Solid-State Circuits Conference, 2006. ESSCIRC 2006. Proceedings of the 32nd European, Digital Object Identifier: 10.1109/ESSCIR.2006.307533 Publication Year: 2006, Page(s): 70 – 73. With permission.)

We also feed I_{o1} as an input to TFT T3. The outputs of TFT T3 can be calculated from Equation 12.6 and are

$$
\begin{aligned}
V_{o3} &= \left(\frac{I_{o1}}{\beta_3}\right)^{1/2} \frac{1}{1-f(t)} \qquad (12.12)\\
&= (\beta_1/\beta_3)^{1/2}(V_{gs} - V_{T0})\\
I_{o3} &= I_{o1}\\
&= \beta(V_{gs} - V_{T0})^2(1 - f(t))^2
\end{aligned}
$$

Thus, in general, all the outputs of all possible connections are always time invariant voltages or time variant currents. *This is exactly like if the entire circuit were responding like a single TFT under constant gate bias.*

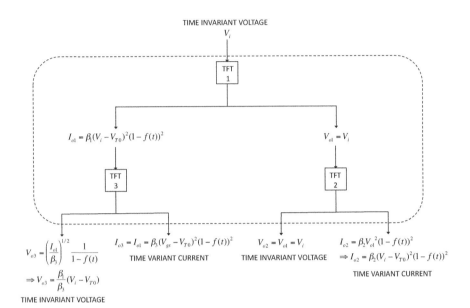

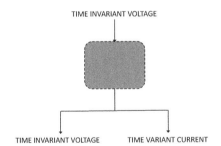

FIGURE 12.4
Circuits of two TFTs with time invariant voltage input. (From Sambandan, S.; Nathan, A.; Circuit Techniques for Organic and Amorphous Semiconductor based Field Effect Transistors, IEEE Solid-State Circuits Conference, 2006. ESSCIRC 2006. Proceedings of the 32nd European, Digital Object Identifier: 10.1109/ESSCIR.2006.307533 Publication Year: 2006, Page(s): 70–73. With permission.)

12.3.2 Current Input

In Figure 12.5 we have the case of TFT T1 being supplied a constant input current, I_{ds}. The outputs of T1 as shown in Figure 1b will be

$$V_{o1} = \left(\frac{I_{ds}}{\beta_1}\right)^{1/2} \frac{1}{1 - f(t)} \qquad (12.13)$$

$$I_{o1} = I_{ds}$$

These outputs of T1 can now be fed to another TFT. We feed V_{o1} as an input to TFT T2. The outputs of T2 then are

$$V_{o2} = V_{o1} = \left(\frac{I_{ds}}{\beta_1}\right)^{1/2} \frac{1}{1 - f(t)} \qquad (12.14)$$

$$I_{o2} = \beta_2 V_{o1}^2 (1 - f(t))^2$$

$$= (\beta_2/\beta_1) I_{ds}$$

We also feed I_{o1} as an input to TFT T3. The outputs of TFT T3 can be calculated from Equation 12.6 and are

$$V_{o3} = \left(\frac{I_{o1}}{\beta_3}\right)^{1/2} \frac{1}{1 - f(t)} \qquad (12.15)$$

$$I_{o3} = I_{o1} = I_{ds}$$

Thus, in general, all the outputs of all possible connections are always time invariant currents or time variant voltage. *This is exactly like if the entire circuit were responding like a single TFT under constant current bias.*

12.4 Paradigm for Circuit Synthesis

Based on the above discussions we see a trend emerging. Whenever the input to a TFT circuit is a time invariant voltage, then we have time invariant voltage outputs and time variant current outputs. On the other hand when the input to the TFT is a time invariant current, we have time invariant current outputs and time invariant voltage outputs.

Thus we establish a simple rule for building circuits with time invariant transfer functions (output to input ratio). We state that *TFT circuits can be made to self compensate for intrinsic VT shift in the TFTs comprising the circuit if — the circuit consists of only TFTs, all of them operating in continous mode, and all of them biased in saturation — and the input and output of the circuit are in the same domain i.e., voltage or current.*

In other words, if we have circuit whose input is a voltage, then the transfer function of the TFT is time invariant if the output is in the voltage domain.

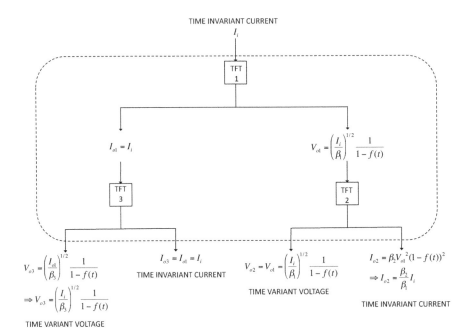

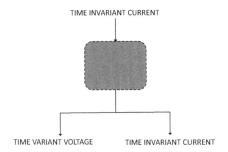

FIGURE 12.5
Circuits of two TFTs with time invariant current input. (From Sambandan, S.; Nathan, A.; Circuit Techniques for Organic and Amorphous Semiconductor based Field Effect Transistors, IEEE Solid-State Circuits Conference, 2006. ESSCIRC 2006. Proceedings of the 32nd European, Digital Object Identifier: 10.1109/ESSCIR.2006.307533 Publication Year: 2006, Page(s): 70–73. With permission.)

If the output is in the current domain, then the transfer function will be time variant. On the other hand if the input to the circuit is current, the transfer function is time invariant if the output is a current and time variant if the output is a voltage.

12.5 Building Blocks

If the above theoretical concept had to be used to design each and every circuit, it would be very cumbersome. Instead, we follow the approach of designing some simple, useful building blocks with the TFT. These building blocks are little circuits which are based on the above concepts and have an inherently time invariant transfer function.

We consider two simple circuits with two TFTs — the voltage amplifier and the current mirror.

12.5.1 Voltage Amplifier

Consider the two TFT common source voltage amplifier as shown in Figure 12.6. TFT T1 acts as a driver transistor and gets the input voltage V_i while TFT T2 acts like a load transistor with a fixed gate bias of V_b. We assume that both TFTs are in saturation and that both TFTs have an initial threshold voltage of V_{T0}. We let the output of the TFT be V_o. In order to find, V_o at any instant in time, the current through T1 can be equated to the current through T2.

12.5.1.1 Time Zero Behavior

The transfer characteristics of this circuit is shown in Figure 12.7. At time zero, where there is no VT shift in the TFTs, The current through T1 is given by

$$I_{d1} = \beta_1 (V_i - V_{T0})^2 \tag{12.16}$$

The current through T2 is given by

$$I_{d2} = \beta_2 (V_b - V_o - V_{T0})^2 \tag{12.17}$$

Equating the currents in T1 and T2, i.e. $I_{d1} = I_{d2}$, we see that

$$V_o = \left(\frac{\beta_1}{\beta_2}\right)^{1/2} V_i - V_b + (1 - \left(\frac{\beta_1}{\beta_2}\right)^{1/2}) V_{T0} \tag{12.18}$$

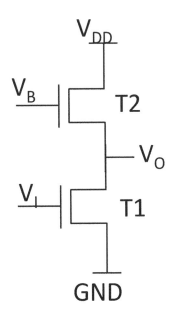

FIGURE 12.6
Two TFT voltage amplifier. The amplifier consists of two TFTs T1 and T2 connected in series between a power supply of V_{DD} and ground. Both TFTs operate in saturation. V_i is the input to the amplifier and V_o is the output. (From Sambandan, S.; Street, R.A.; Self-Stabilization in Amorphous Silicon Circuits, IEEE Electron Device Letters, IEEE, Volume: 30, Issue: 1 Digital Object Identifier: 10.1109/LED.2008.2009010 Publication Year: 2009, Page(s): 45–47. With permission.)

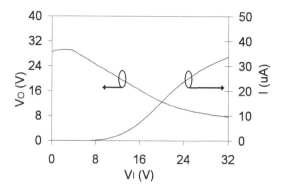

FIGURE 12.7
Transfer characteristics and current voltage characteristics of the voltage amplifier. (From Sambandan, S.; Street, R.A.; Self-Stabilization in Amorphous Silicon Circuits, IEEE Electron Device Letters, IEEE, Volume: 30, Issue: 1 Digital Object Identifier: 10.1109/LED.2008.2009010 Publication Year: 2009, Page(s): 45–47. With permission.)

12.5.1.2 Evolution with Time

Now let us consider the threshold voltage shift with time. T1 receives a constant gate-source bias, while T2 has a current $I_{d2} = I_{d1}$ defined by T1 flowing through it. The threshold voltage shift in TFT T1 and T2 are given by

$$\delta V_{T-T1} = (V_i - V_{T0})f(t) \tag{12.19}$$

$$\delta V_{T-T2} = \left(\frac{I_{d1}}{\beta_2}\right)^{1/2} \frac{f(t)}{1 - f(t)}$$

The gate-source voltage of T2, $V_b - V_o$ is then given by

$$V_b - V_o = \left(\frac{I_{d1}}{\beta_2}\right)^{1/2} \frac{1}{1 - f(t)} + V_{T0} \tag{12.20}$$

However, since $I_{d1} = \beta_1(V_i - V_{T0} - \delta V_{T-T1})^2$, we have

$$V_o = \left(\frac{\beta_1}{\beta_2}\right)^{1/2} V_i - V_b + (1 - \left(\frac{\beta_1}{\beta_2}\right)^{1/2})V_{T0} \tag{12.21}$$

This implies that the threshold voltage shift in the TFT with time does not affect the output voltage.

There is another way to look at this. We can also consider the threshold voltage shift in T2 as given by $(V_b - V_o - V_{T0})f(t)$, where V_o is not neccesarily time invariant, due to the equivalence of current and voltage bias. Thus, the current through T1 at some given time is

$$
\begin{aligned}
I_{d1} &= \beta_1(V_i - V_{T0} - (V_i - V_{T0})f(t))^2 \\
&= \beta_1(V_i - V_{T0})^2(1 - f(t))^2
\end{aligned}
\tag{12.22}
$$

The current through T2 is given by

$$
\begin{aligned}
I_{d2} &= \beta_2(V_b - V_o - V_{T0} - (V_b - V_o - V_{T0})f(t))^2 \\
&= \beta_2(V_b - V_o - V_{T0})^2(1 - f(t))^2
\end{aligned}
\tag{12.23}
$$

Equating the two current, we see that

$$
V_o = \left(\frac{\beta_1}{\beta_2}\right)^{1/2} V_i - V_b + (1 - \left(\frac{\beta_1}{\beta_2}\right)^{1/2})V_{T0}
\tag{12.24}
$$

which is the same answer as obtained above.

12.5.1.3   Implications

If we now refer back to the stability paradigm, we see that since the circuit consists of only TFTs, and all TFTs are operating in saturation, the transfer characteristics or transfer function of the circuit, $\frac{V_o}{V_i}$, is indeed time invariant, in spite of the threshold voltage shift in the TFTs. TFT T1 receives a constant input voltage V_i and results in a time varying current. This time varying current is sourced from TFT T2, and therefore the output voltage, which is related to the gate-source voltage of T2, is constant.

12.5.1.4   Experimental Verifications

Experimental results shown in Figure 12.8, Figure 12.9 and Figure 12.10 agree with the predictions of the theory. Figure 12.8 shows the output voltage of the voltage amplifier for different input voltages. The experiment is performed on two voltage amplifiers with gain parameter $A = 1$ and $A = 2$. Figure 12.9 shows that the threshold voltage in the TFT T2 is changing with time. In spite of this change, the output of the amplifier is time invariant as long as T1 and T2 are in saturation (Figure 12.8). Figure 12.10 illustrates the change in the transfer characteristics of the amplifier before and after the stress test. Also shown is the change in the amplfier current. It is seen that the current change is significant due to the threshold voltage shift. Thus the amplifier maintained a constant transfer characteristics (output to input votage ratio) at the cost of a time variant internal variable (current through the circuit).

12.5.1.5   Time Varying Input Voltage

We now consider the case when the input voltage to the amplfier, V_i is varying with time. We split the time interval into units of length τ. Let the sequence of inputs to the amplifier be $V_{i,1}, V_{i,2} \cdots V_{i,n}$. Note that the output voltage

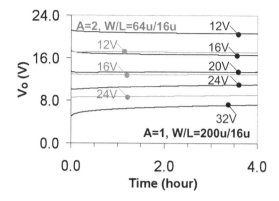

FIGURE 12.8
Output voltage of the 2 TFT voltage amplifier for different input voltages. The parameter $A = (\beta_1/\beta_2)^{1/2}$. The behaviour of two amplifiers with $A = 1$ and $A = 2$ are shown. (From Sambandan, S.; Street, R.A.; Self-Stabilization in Amorphous Silicon Circuits, IEEE Electron Device Letters, IEEE, Volume: 30, Issue: 1 Digital Object Identifier: 10.1109/LED.2008.2009010 Publication Year: 2009, Page(s): 45. With permission.)

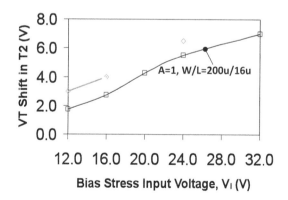

FIGURE 12.9
Threshold voltage shift in T2 before and after the stress test. (From Sambandan, S.; Street, R.A.; Self-Stabilization in Amorphous Silicon Circuits, IEEE Electron Device Letters, IEEE, Volume: 30, Issue: 1 Digital Object Identifier: 10.1109/LED.2008.2009010 ?Publication Year: 2009, Page(s): 45. With permission.)

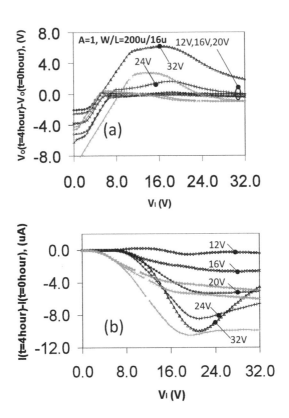

FIGURE 12.10
(a) The difference between the transfer characteristics of the amplifier before and after the stress test. (b) The difference in the amplifier current characteristics before and after the stress test. (From Sambandan, S.; Street, R.A.; Self-Stabilization in Amorphous Silicon Circuits, IEEE Electron Device Letters, IEEE, Volume: 30, Issue: 1 Digital Object Identifier: 10.1109/LED.2008.2009010 Publication Year: 2009, Page(s): 45. With permission.)

will definitely be time dependent since the inputs are changing, but we can show that the output voltage will not be dependent on the threshold voltage shift. Let the sequence of output voltages corresponding to the above sequence of input voltages be $V_{o,1}, V_{o,2}, \cdots V_{o,n}$. The master equation governing the behavior of this circuit at any time interval $(j-1)\tau \leq t < j\tau$ is that the current in both the TFTs are the same and hence

$$\beta_1(V_{i,j} - V_{T,j-1}^{T1})^2 = \beta_2(V_b - V_{o,j} - V_{T,j-1}^{T2})^2 \tag{12.25}$$

where $V_{T,j-1}^{T1}$ and $V_{T,j-1}^{T2}$ are the threshold voltages of T1 and T2 at the end of the previous time interval.

As seen earlier, at time zero and therefore in the first time interval there is no threshold voltage shift and

$$V_{o,1} = V_b - (\beta_1/\beta_2)^{1/2}V_{i,1} + ((\beta_1/\beta_2)^{1/2} - 1)V_{T0} \tag{12.26}$$

According to Equation 12.8 and Equation 26, the threshold voltage shift in T1 and T2 at the end of the first time interval are given by

$$
\begin{aligned}
V_{T,1}^{T1} &= V_{T0} + \varphi f(\tau)(V_{i,1} - V_{T0}) \\
V_{T,1}^{T2} &= V_{T0} + \varphi f(\tau)(V_b - V_{o,1} - V_{T0}) \\
&= V_{T0} + (\beta_1/\beta_2)^{1/2}\varphi f(\tau)(V_{i,1} - V_{T0})
\end{aligned}
\tag{12.27}
$$

Therefore $(\beta_1/\beta_2)^{1/2}(V_{T,1}^{T1} - V_{T0}) = (V_{T,1}^{T2} - V_{T0})$. In the second time interval, using Equation 27 in Equation 12.25 and equating the currents in T1 and T2 we see that

$$V_{o,2} = V_b - (\beta_1/\beta_2)^{1/2}V_{i,2} + ((\beta_1/\beta_2)^{1/2} - 1)V_{T0} \tag{12.28}$$

In other words, the output voltage did not change with the threshold voltage shift.

We can now show this to be true for the time interval $(n-1)\tau \leq t < n\tau$ using the *Principle of Mathematical Induction*. Let the result be true for the interval $(n-2)\tau \leq t < (n-1)\tau$. Therefore $(\beta_1/\beta_2)^{1/2}(V_{T,n-1}^{T1} - V_{T0}) = (V_{T,n-1}^{T2} - V_{T0})$ and

$$V_{o,n-1} = V_b - (\beta_1/\beta_2)^{1/2}V_{i,n-1} + ((\beta_1/\beta_2)^{1/2} - 1)V_{T0} \tag{12.29}$$

The cumulative threshold voltages in TFTs T1 and T2 at the end of the interval $(n-1)\tau \le t < n\tau$ are given by Equation 12.7 to be

$$
\begin{aligned}
V_{T,n}^{T1} &= V_{T0} + \varphi f(\tau) \sum_{j=0}^{j=n} (V_{i,j} - V_{T0})(1 - f(\tau))^{n-j} \qquad (12.30)\\
&= V_{T,n-1}^{T1} + \varphi f(\tau)(V_{i,n} - V_{T0})\\
V_{T,n}^{T2} &= V_{T0} + \varphi f(\tau) \sum_{j=0}^{j=n} (V_b - V_{o,j} - V_{T0})(1 - f(\tau))^{n-j}\\
&= V_{T,n-1}^{T2} + \varphi f(\tau)(V_b - V_{o,n} - V_{T0})\\
&= (\beta_1/\beta_2)^{1/2} V_{T,n-1}^{T1} + \varphi f(\tau)(V_b - V_{o,n} - V_{T0})
\end{aligned}
$$

Equating the currents in T1 and T2 using the master equation Equation 12.25 and using Equation 12.30, we have

$$
V_{o,n} = V_b - (\beta_1/\beta_2)^{1/2} V_{i,n} + ((\beta_1/\beta_2)^{1/2} - 1) V_{T0} \qquad (12.31)
$$

This implies that even for a time varying input bias, the output voltage does not depend on the threshold voltage shift.

12.5.2   Current Mirror

Next let us consider the current mirror which in some sense is also a current amplifier. The current mirror consists of two transistors as shown in Figure 12.11. TFT T1 is diode connected and receives a constant input current I_i. TFT T2 shares the same gate voltage as T1 and sinks an output current I_o. Both T1 and T2 are biased in saturation and share the same gate to source voltage due to the topology of the circuit. We assume that both TFTs have the same initial threshold voltage, V_{T0}.

12.5.2.1   Time Zero Behavior

The transfer characteristics of the current mirror is shown in Figure 12.12. At time zero, we have the input current into TFT T1 forcing the gate-source voltage to go to V_{g1} where

$$
V_{g1} - V_{T0} = \frac{I_i}{\beta_1} \qquad (12.32)
$$

Since the gate to source voltage of T2 is the same as that of T1, i.e. $V_{g2} = V_{g1}$, we have the output current, I_o related to the input current I_i as

$$
\begin{aligned}
I_o &= \beta_2 (V_{g2} - V_{T0})^2 \qquad (12.33)\\
&= \frac{\beta_2}{\beta_1} I_i
\end{aligned}
$$

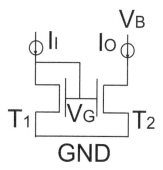

FIGURE 12.11

Two TFT current mirror. The current mirror consists of two TFTs T1 and T2 connected with a common gate. Both TFTs operate in saturation. I_i is the input to the amplifier and I_o is the output. (From Sambandan, S.; Street, R.A.; Self-Stabilization in Amorphous Silicon Circuits, IEEE Electron Device Letters, IEEE, Volume: 30, Issue: 1 Digital Object Identifier: 10.1109/LED.2008.2009010 Publication Year: 2009, Page(s): 45. With permission.)

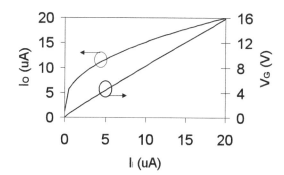

FIGURE 12.12

Transfer characteristics and current voltage characteristics of the current mirror. (From Sambandan, S.; Street, R.A.; Self-Stabilization in Amorphous Silicon Circuits, IEEE Electron Device Letters, IEEE, Volume: 30, Issue: 1 Digital Object Identifier: 10.1109/LED.2008.2009010 Publication Year: 2009, Page(s): 45. With permission.)

12.5.2.2   Evolution with Time

With time, there occurs a threshold voltage shift in both T1 and T2. The threshold voltage shift in T1 is due to the input current, while the threshold voltage shift in T2 is due to the presence of the gate bias of T2 which is the same as the gate bias of T1. Thus, the threshold voltage shift in T1 and T2 are given by

$$\delta V_{T-T1} = \left(\frac{I_i}{\beta_1}\right)^{1/2} \frac{f(t)}{1 - f(t)} \tag{12.34}$$

$$\delta V_{T-T2} = (V_{g1} - V_{T0})f(t)$$

However, since $V_{g1} - V_{T0} - \delta V_{T-T1} = \frac{I_i}{\beta_1}$, the output current, is therefore given by

$$I_o = \beta_2(V_{g2} - V_{T0} - (V_{g1} - V_{T0})f(t))^2 \tag{12.35}$$

$$= \frac{\beta_2}{\beta_1} I_i$$

That is, the output current is also time invariant.

12.5.2.3   Implications

Once again looking at the stability paradigm, we see that since the circuit consists of only TFTs, and all TFTs are operating in saturation, the transfer characteristics or transfer function of the circuit, $\frac{I_o}{I_i}$, is indeed time invariant, in spite of the threshold voltage shift in the TFTs. Using the modular representation of the TFT, Figure 6b describes what is happening conceptually. TFT T1 receives a constant input current I_i and results in a time varying gate-source bias which attempts to adjust its value to compensate for the threshold voltage shift. This time varying voltage is supplied to the gate of TFT T2. Since the gate-source voltage in T1 and T2 are the same, and the gate-source voltage is changing with time trying to compensate for the threshold voltage shift, the output current, which is related to the gate-source voltage of T2, is constant.

12.5.3   Experiments

Experimental results shown in Figure 12.13, Figure 12.14 and Figure 12.15 agree with the predictions of the theory. Figure 12.13 shows the output current of the current mirror (or current amplifier) for different input currents. The experiment is performed on two current amplifiers with gain parameter $A = 1$ and $A = 2$. Figure 12.14 shows that the threshold voltage in the TFT T2 is changing with time. In spite of this change, the output of the current amplifier is time invariant as long as T1 and T2 are in saturation (Figure 12.13). Figure 12.15 illustrates the change in the transfer characteristics of the current mirror before and after the stress test. Also shown is the change in the gate voltage

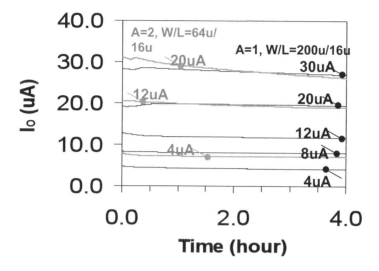

FIGURE 12.13
Output current of the 2 TFT current mirror for different input currents during stress. The parameter $A = (\beta_2/\beta_1)^{1/2}$. The behaviour of two current mirrors with $A = 1$ and $A = 2$ are shown. (From Sambandan, S.; Street, R.A.; Self-Stabilization in Amorphous Silicon Circuits, IEEE Electron Device Letters, IEEE, Volume: 30, Issue: 1 Digital Object Identifier: 10.1109/LED.2008.2009010 Publication Year: 2009, Page(s): 45. With permission.)

developed at the gate of TFT T2. It is seen that the gate voltage change is significant due to the threshold voltage shift. Thus the amplifier maintained a constant transfer characteristics (output to input current ratio) at the cost of a time variant internal variable (gate voltage).

12.5.4 Time Varying Input Current

We now consider the case when the input voltage to the current mirror, I_i is varying with time. We split the time interval into units of length τ. Let the sequence of inputs to the amplifier be $I_{i,1}, I_{i,2} \cdots I_{i,n}$. The output current will be time dependent since the inputs are changing, but we can show that the output current will not be dependent on the threshold voltage shift. Let the sequence of output currents corresponding to the above sequence of input currents be $I_{o,1}, I_{o,2}, \cdots I_{o,n}$. The master equation governing the behavior of this circuit at any time interval $(j-1)\tau \le t < j\tau$ is that the gate voltage on

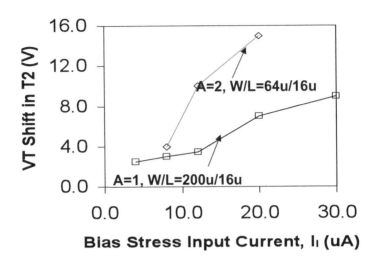

FIGURE 12.14
Threshold voltage shift in T2 before and after the test described in Figure 12.13. (From Sambandan, S.; Street, R.A.; Self-Stabilization in Amorphous Silicon Circuits, IEEE Electron Device Letters, IEEE, Volume: 30, Issue: 1 Digital Object Identifier: 10.1109/LED.2008.2009010 Publication Year: 2009, Page(s): 45. With permission.)

both the TFTs are the same and hence

$$(I_{i,j}/\beta_1)^{1/2} + V^{T1}_{T,j-1} = (I_{o,j}/\beta_2)^{1/2} + V^{T2}_{T,j-1} \tag{12.36}$$

where $V^{T1}_{T,j-1}$ and $V^{T2}_{T,j-1}$ are the threshold voltages of T1 and T2 at the end of the previous time interval.

As seen earlier, at time zero and therefore in the first time interval there is no threshold voltage shift and

$$I_{o,1} = (\beta_2/\beta_1)I_{i,1} \tag{12.37}$$

According to Equation 12.4, the threshold voltage shift in T1 and T2 at the end of the first time interval are given by

$$
\begin{aligned}
V^{T1}_{T,1} &= V_{T0} + \frac{\varphi f(\tau)}{1 - \varphi f(\tau)}(I_{i,1}/\beta_1)^{1/2} \tag{12.38}\\
V^{T2}_{T,1} &= V_{T0} + \frac{\varphi f(\tau)}{1 - \varphi f(\tau)}(I_{o,1}/\beta_2)^{1/2}\\
&= V_{T0} + \frac{\varphi f(\tau)}{1 - \varphi f(\tau)}(I_{i,1}/\beta_1)^{1/2}
\end{aligned}
$$

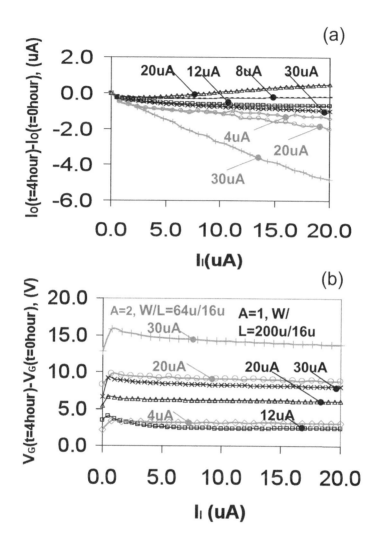

FIGURE 12.15

(a) The difference between the transfer characteristics of the current mirror before and after the stress test. (b) The difference in the current mirror gate voltage characteristics before and after the stress test. (From Sambandan, S.; Street, R.A.; Self-Stabilization in Amorphous Silicon Circuits, IEEE Electron Device Letters, IEEE, Volume: 30, Issue: 1 Digital Object Identifier: 10.1109/LED.2008.2009010 Publication Year: 2009, Page(s): 45. With permission.)

Thus, $V_{T,1}^{T1} = V_{T,2}^{T2}$. In the second time interval, substituting for these values of threshold voltage in the master equation Equation 12.36

$$I_{o,2} = (\beta_2/\beta_1)I_{i,2} \tag{12.39}$$

In other words, the output current does not depend on the threshold voltage shift.

We can now show this to be true for the time interval $(n-1)\tau \le t < n\tau$ using the Principle of Mathematical Induction. Let the result be true for the interval $(n-2)\tau \le t < (n-1)\tau$. Therefore

$$
\begin{aligned}
V_{T,n-1}^{T1} &= V_{T,n-1}^{T2} & (12.40) \\
I_{o,n-1} &= (\beta_2/\beta_1)^{1/2} I_{i,n-1}
\end{aligned}
$$

The cumulative threshold voltages in TFTs T1 and T2 are given by Equation 12.9 to be

$$
\begin{aligned}
V_{T,n}^{T1} &= V_{T0} + \frac{\varphi f(\tau)}{1 - \varphi f(\tau)} \sum_{j=1}^{n} \left(\frac{I_{i,j}}{\beta_1}\right)^{1/2} \left(\frac{1 - f(\tau)}{1 - \varphi f(\tau)}\right)^{n-j} & (12.41) \\
&= V_{T,n-1}^{T1} + \frac{\varphi f(\tau)}{1 - \varphi f(\tau)} \left(\frac{I_{i,n}}{\beta_1}\right)^{1/2} \\
V_{T,n}^{T2} &= V_{T0} + \frac{\varphi f(\tau)}{1 - \varphi f(\tau)} \sum_{j=1}^{n} \left(\frac{I_{o,j}}{\beta}\right)^{1/2} \left(\frac{1 - f(\tau)}{1 - \varphi f(\tau)}\right)^{n-j} \\
&= V_{T,n-1}^{T2} + \frac{\varphi f(\tau)}{1 - \varphi f(\tau)} \left(\frac{I_{o,n}}{\beta_2}\right)^{1/2} \\
&= V_{T,n-1}^{T1} + \frac{\varphi f(\tau)}{1 - \varphi f(\tau)} \left(\frac{I_{o,n}}{\beta_2}\right)^{1/2}
\end{aligned}
$$

Substituting for $V_{T,n}^{T1}$ and $V_{T,n}^{T2}$ in Equation 12.36, we obtain

$$I_{o,n-1} = (\beta_2/\beta_1)^{1/2} I_{i,n-1} \tag{12.42}$$

Therefore, we prove that in spite of the threshold voltage shift in the TFTs of the current mirror under the condition of a time varying input bias, the transfer characteristics of the current mirror are not dependent on the threshold voltage shift.

12.6   Extending the Design Paradigm

In the previous section, we considered two useful building blocks for integrated circuit design with the TFT. In the next chapter we make use of the building blocks to design a high gain voltage amplifier with TFTs which show time invariant behaviour. In this section, we extend the concept of circuit design - the design paradigm - to some generalisations of circuits. In particular we consider (without mathematical proofs) the cascoded voltage amplifier which is a generalisation of the two TFT voltage amplifier, the translinear circuits with a stacked loop topology which are a generalisation of the current mirror, and the concept of feedback in circuits.

12.6.1 Cascode Voltage Amplifiers

While we discussed the common source voltage amplifier in the previous section, it must be noted that all circuits with that topology — e.g., common drain amplifier and common gate amplifiers, will also show the time invariant transfer characteristics to threshold voltage shift in the TFTs.

A *cascode amplifier* is typically referred to a stack comprising of a common source and common gate amplifier for very high gain. The amplifier circuit is shown in Figure 12.16a.

In this discussion we refer to a cascode stack of TFTs connected in series as shown in Figure 12.16b. Clearly, the current through all the TFTs must be the same. If all the TFTs operate in saturation, we have (for some TFT j in the stack)

$$V_{bias-j} - V_{out-j+1} - V_{T0} = constant/\beta_j \qquad (12.43)$$

Thus, the voltage at every drain node of the TFTs will generally be time invariant for constant gate bias conditions as long as the TFTs are biased in saturation mode operation.

12.6.2 Translinear Circuits

The translinear circuit is shown in Figure 12.17 and is a generalisation of the current mirror [184]. Since the top two TFTs share the same gate voltage, the following clockwise and anti-clockwise summation must hold true

$$\sum_{j=1}^{M} \left[I_d^{Aj}/\beta^{Aj} + V_T^{Aj} \right] = \sum_{j=1}^{M} \left[I_d^{Bj}/\beta^{Bj} + V_T^{Bj} \right]. \qquad (12.44)$$

It can be shown that the translinear circuit also has a property of having a threshold voltage shift independent behavior as long as the TFTs remain in saturation.

12.6.3 Feedback Loops

The above discussions were concerned with open structures where the flow of information did not loop back or cross paths. But what if we have feedback loops, for example, what if the output of T3 were feedback to input of T2?

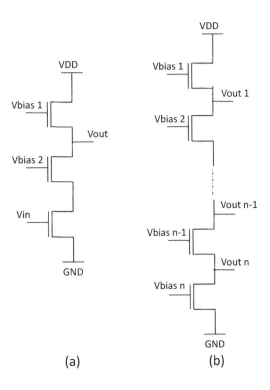

(a)   (b)

FIGURE 12.16
(a) Cascode amplifier. (b) Voltage amplifier comprised of a stack of TFTs
corrected in series and operating in saturation mode.

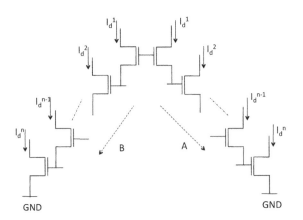

FIGURE 12.17
Translinear circuit with a stacked topology.

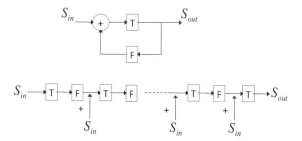

FIGURE 12.18
Circuits with open loop transfer function T and feedback transfer function F.

Let us consider a TFT circuit with transfer function T having a feedback F as shown in Figure 12.18. Let the input to the circuit be S_{in}, which may either be in voltage or current mode, and the output of the circuit be S_{out}, which is also either voltage or current. We impose a constraint on our feedback loop that $|TF| < 1$.

Based on Figure 12.18, we see that $S_{out} = T(S_{in} + FS_{out})$, and therefore the effective transfer function of the circuit with feedback is

$$\frac{S_{out}}{S_{in}} = \frac{T}{1 - TF} \tag{12.45}$$

What can we say about the stability paradigm for circuits with feedback? It appears that our simple graphs of open tree like structures as shown in Figure 12.2 need to be made more complex.

However, there is another way of visualizing feedback in circuits. Since the signal in a feedback loop goes round and round the loop, we can open the loop into an infinitely long open path as shown in Figure 12.18. As we traverse along this path we find that S_{out} is given by

$$
\begin{aligned}
S_{out} &= ((\dots((S_{in}TF + S_{in})TF + \dots + S_{in})TF + S_{in})T \quad (12.46) \\
&= S_{in}(TF + (TF)^2 + (TF)^3 + \dots (TF)^\infty)\frac{1}{F} \\
&= S_{in}\frac{T}{1 - TF}
\end{aligned}
$$

Thus the transfer function of the open circuit representation of the feedback loop results in the same transfer function as the closed loop of Figure 12.18.

We have shown that open tree like circuits can have time invariant transfer characterisitcs when the input and output are in the same domain. Thus, the same is applicable even to feedback loops as feedback loops can be opened out to infinite stage open circuits.

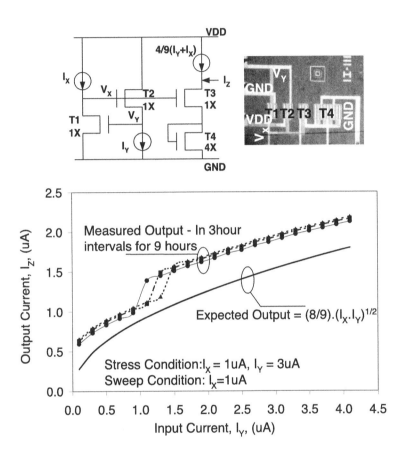

FIGURE 12.19
Proof of concept using a root mean square estimation circuit. (From Samban-dan, S.; Nathan, A.; Circuit Techniques for Organic and Amorphous Semiconductor based Field Effect Transistors, IEEE, Solid-State Circuits Conference, 2006. ESSCIRC 2006. Proceedings of the 32nd European, Digital Object Identifier: 10.1109/ESSCIR.2006.307533, Publication Year: 2006, Page(s): 70–73. With permission.)

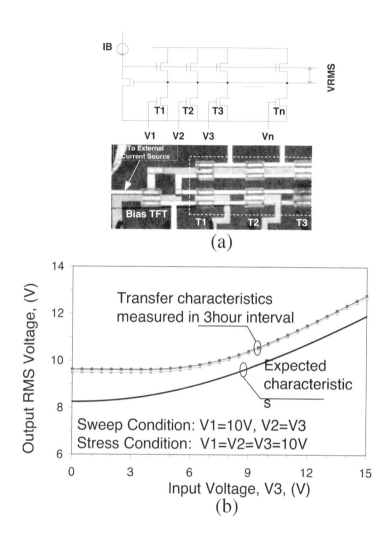

FIGURE 12.20
Proof of concept using a geometric mean circuit. (From Sambandan, S.; Nathan, A.; Circuit Techniques for Organic and Amorphous Semiconductor based Field Effect Transistors, IEEE, Solid-State Circuits Conference, 2006. ESSCIRC 2006. Proceedings of the 32nd European, Digital Object Identifier: 10.1109/ESSCIR.2006.307533, Publication Year: 2006, Page(s): 70–73. With permission.)

12.7   Examples

We illustrate as a proof of concept, a circuit to estimate the root mean square of an input voltage vector. Figure12.19a illustrates the root mean square voltage generation circuit. The circuit is modular with each voltage amplifier being a module that accepts an input voltage. An n module circuit accepts the input voltage vector, $V1, V2, \cdots Vn$ and returns their root mean square value. Figure 12.19b illustrates the time invariant transfer characteristics of the root mean square circuit in spite of the V_T shift.

Figure 12.20 illustrates a weighted geometric mean circuit whose output is the weighted geometric mean of the two inputs, and in this case $I_{out} = 8\sqrt{I_x I_y}/9$ along with time invariance of the output. This circuit uses the translinear design principle and it is seen that the transfer characteristics are time invariant.

12.8   Conclusion

In this chapter we discussed an interesting approach of designing circuits based on non-crystalline semiconductors with the aim of offsetting the problems created by the threshold voltage shift in the TFTs. We found that by using TFTs in certain topologies, the transfer characteristics of the circuit could be made immune to the threshold voltage shift in the individual TFTs. In the next chapter we see a definite application of this idea towards the design of high gain amplifiers.

13

Case Study — Pseudo PMOS Field Effect
Transistor

CONTENTS

In many disordered semiconductor materials the effective mobility of one specie of carriers (electrons/holes) is much higher than the other(holes/electrons). For example, in the case of hydrogenated amorphous silicon based TFTs, the electron mobility is significantly higher than the hole mobility. This leads to TFTs of n-type being much stronger than p-type. The case is opposite in many polymer semiconductors such as poly(3,3 dialkylquater-thiophene). Therefore, electronics based on such semiconductors are for all practical purposes devoid of complementary devices.

The aim of this chapter is to show how one can design a high gain amplifier and biasing circuits such as current sinks and sources using a non complementary process with the aid of feedback. The highlight is the design of a current source with n-type TFTs (or equivalently, a current sink with p-type TFTs) using positive feedback. The high impedance of the current source is then used to build high gain amplifiers with a non-complementary process [141].

13.1   Role of Complementary Devices

13.1.1   TFT as Current Sources and Sinks

Let us revisit the output characteristics of a n-type and a p-type TFT biased at a certain $|V_{GS}| > |V_T|$. Figure 13.1a and Figure 13.1b shows the output characteristics of a n-type with $V_{GS} = 5V$ and a p-type TFT with $V_{GS} = -5V$, respectively. We see that the TFTs behave like a finite resistor of low impedance (and equal to $\partial V_{DS}/\partial I_{DS}$) till just before saturation. Beyond a certain V_{DS}, the current saturates. In this region the n-type TFT behaves like an almost ideal *current sink*, where it sinks to ground (GND) an almost constant level of current even if V_{DS} is increased. On the other hand, the saturated p-type TFT behaves like an almost ideal *current source*, where it sources from power supply an almost constant level of current even if V_{DS} is decreased.

The impedance of the TFT is given by $\partial V_{DS}/\partial I_{DS}$. If the saturation region were to be flat, and I_{DS} did not depend on V_{DS} at all, then the impedance would be infinite. In other words, when seen from the drain terminal, the TFTs have a high impedance after saturation. In reality, and as seen in Figure 13.1, the output curves of both n-type and p-type TFTs do not perfectly saturate, i.e., there exists a small dependence of I_{DS} on the drain-source voltage, V_{DS} even in the saturation region. In other words, the n-type TFT is not a perfect current source and the p-type TFT is not a perfect current sink even in saturation. We designate this finite impedance of the n-type and p-type TFT in saturation as r_{on} and r_{op}, respectively.

While the terms *current source* and *current sink* are in a way interchangeable - since a current sink sources negative current - the use of these terms aids understanding in this chapter. It is thus important to understand this difference in the n-type and p-type TFT. As shown in the circuit of Figure 13.2a, the n-type TFT in saturation draws and sinks (to ground) a constant current from any arbitrary circuit element. On the other hand, as shown in Figure 13.2b, the p-type TFT in saturation sources (from supply) a constant current to any arbitrary circuit element.

13.1.2   Benefits of a Complementary Device

In the short discussion above we saw that p-type TFT can be a good current source while an n-type TFT can be a good current sink when in the saturation mode of operation. As a good current source and sink, they have a high output impedance when seen from the drain terminal. This proves to be a major advantage when both n-type and p-type are used to complement each other in circuit design.

In order to understand the benefits of the high output impedance provided by the complementary TFTs we study a common source amplifier with a n-

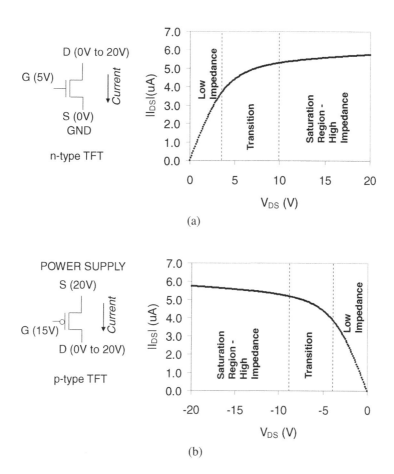

FIGURE 13.1

(a) Output characteristics of a n-type TFT. After saturation the n-type TFT behaves like a good current sink with a high impedance of r_{on} seen from the drain terminal. (b) Output characteristics of a p-type TFT. After saturation the p-type TFT behaves like a good current source with a high impedance of r_{op} seen from the drain terminal.

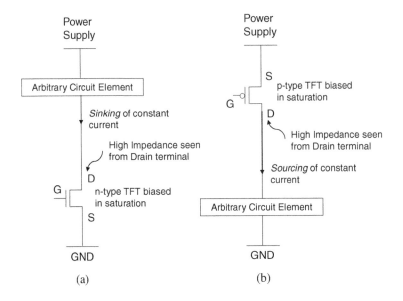

FIGURE 13.2
(a) n-type TFT as a current sink (b) p-type TFT as a current source.

type TFT driver. Our goal is to design this amplifier for maximum dc-gain by considering different options for the load element.

13.1.2.1   n-type TFT Load

Let us consider a n-type TFT load as shown in Figure 13.3a with both TFTs biased in saturation. The amplifier has an n-type TFT load connected between the supply voltage, V_{DD} and the drain terminal of the n-type driver TFT. The input voltage to the amplifier, V_i is presented at the gate of the driver TFT and the output voltage is V_o.

Using the small signal model discussed in the earlier chapters we see that the dc gain of the amplifier is $A = -g_{mD} z_{out}$, where g_{mD} is the transconductance of the driver TFT, and z_{out} the effective output impedance. The effective output impedance, z_{out} is the resistance of the parallel combination of the load impedance and driver. The impedance of the driver TFT seen from the output node is r_{on}, but what is the impedance of the load TFT? If there was a small change in the output voltage, δV_o, the current through the load TFT would vary by $\delta I = -g_{mL} V_o$, where g_{mL} is the transconductance of the load TFT. Thus the effective impedance of the load TFT is $|\delta V_o / \delta I| = 1/g_{mL}$.

The dc gain of the amplifier is thus given by $A = -g_{mD} \frac{r_{on}/g_{mL}}{r_{on} + 1/g_{mL}}$. If $1/g_{mL} \ll r_o$, which is usually the case, $A = -g_{mD}/g_{mL} \propto \sqrt{\frac{W_D/L_D}{W_L/L_L}}$ where

W_D and W_L are the channel width of the driver and load TFT, respectively and L_D and L_L are the channel lengths of the driver and load TFT, respectively. The gain of the amplifier is varied by varying the geometry of the driver and load TFT.

Figure 13.3b illustrates the transfer characteristics of the common source amplifier with n-type TFT load. The curves labeled A, B, C, correspond to the load TFT being in saturation with $V_B = V_{DD} = 20V$ and the TFTs having the following aspect ratios - A: W_D/L_D= 50um/10um, W_L/L_L=50um/10um; B: W_D/L_D= 100um/10um, W_L/L_L=10um/100um; C: W_D/L_D= 5000um/10um, W_L/L_L=10um/10um. Curve D corresponds to the load TFT biased in linear mode of operation and behaving as a resistors with $V_B = 40V$ and with W_D/L_D= 100um/10um, W_L/L_L=10um/100um. Figure 13.3c shows the dc-gain of the amplifier and is obtained by extracting $\partial V_o/\partial V_i$ from Figure 13.3b. The labels of the curves correspond to the labels in Figure 13.3b.

The gain is easily modulated by scaling the aspect ratio of the load TFT. However, the problem of using a n-type TFT load is a tradeoff between large devices sizes and gain. The reason for poor gain is the impedance of the n-type TFT load. When seen from the output node the load TFT has an absolute impedance of $1/g_{mL}$ because changes in the voltage at the output node, V_o, causes a variation in load TFT current. This change in current is due to the fact that the TFT is n-type.

What would we need as a load element if variations on the output node voltage, V_o, should not cause variations in the load current? What about a p-type TFT?

13.1.2.2   p-type TFT Load

Consider the amplifier in Figure 13.4. We saw that if the p-type TFT is biased in saturation, the impedance of the TFT as seen from its drain terminal, which is also the output node of the amplifier is high and equal to r_{op}.

The effective impedance of the amplifier as seen from the output node is the parallel combination of r_{op} and r_{on} since both TFTs are in saturation. The gain of the amplifier in this case is then given by $g_{mD} \frac{r_{op}r_{on}}{r_{op}+r_{on}}$. Since both r_{op} and r_{on} are relatively large, the gain too is large.

It is not required to have transistors of very large sizes in order to obtain this gain. The p-type TFT behaves like an ideal current source with high output impedance to boost gain. *This in essence is the value of a current source load achieved with a complementary device.*

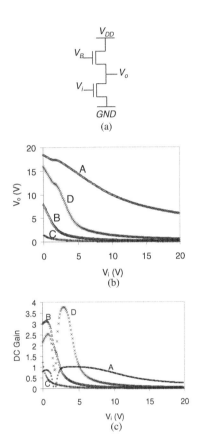

FIGURE 13.3

(a) Common source amplifier with a n-type TFT driver and n-type TFT load. (b) Transfer characteristics of the common source amplifier for different geometries. (c) DC Gain of the amplifer $(\partial V_o / \partial V_i)$ extracted from Figure 3b.

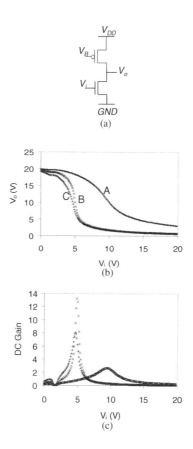

FIGURE 13.4

(a) Common source amplifier with a n-type TFT driver and p-type TFT load. (b) Transfer characteristics of the common source amplifier for different geometries. (c) DC Gain of the amplifer $(\partial V_o / \partial V_i)$ extracted from Figure 13.4b.

13.2   High Impedance Load with a Non-Complementary Process

In the previous section we saw the benefits of high output impedance in the load transistor by comparing different types of amplifiers. It was seen that the complementary device plays a significant role particularly because of this feature. This has significant impact in both digital and analog circuit design. It is the presence of complementary transistors that make a chain of digital logic blocks retain information with less noise. In analog circuit design the presence of complementary devices provides for both current sinks and current sources aiding in efficient methods of biasing circuits. Moreover, the high gain provided by the use of complementary devices in amplifiers is of use in the efficient design of operational amplifiers.

Since many of the disordered semiconducting materials used in TFTs have predominantly electron (or hole) transport, access to complementary devices is not practical. Without the presence of complementary devices, the above benefits cannot manifest in circuits.

In this section, using amorphous silicon as the semiconducting material, we design a circuit that uses TFT of one type (n-type) to create a high impedance load. The primary characteristics we need in the load is that it behave close to an ideal current source — like a p-type TFT. In other words, when the circuit is properly biased, the output impedance of the circuit must be high. This implies that the current sourced by the circuit, I_S as illustrated in Figure 13.5, must be invariant of changes in voltage, V_o on the output node.

13.2.1   Design of the High Impedance Load

13.2.1.1   The Concept

We begin the design by studying the n-type TFT with the drain terminal connected to the power supply and the source terminal sourcing current I_S into some arbitrary circuit element as shown in Figure 13.6a We define the source terminal of the TFT as the output node which has a potential, V_o. Let the n-type TFT be biased with a constant gate voltage, V_B, which biases the TFT in saturation. We see that variations in V_o causes variations in I_S since the gate-source voltage of the n-type TFT varies. In fact,

$$I_S = \beta_n (V_B - V_o - v_{Tn})^2 \qquad (13.1)$$

where v_{Tn} is the threshold voltage of the n-type TFT.

In order to achieve a high output impedance, I_S must not vary with V_o, which is to say that the TFT must behave like a current source. This can be achieved only if the gate-source voltage is kept constant, which in turn is achieved if *the gate voltage follows the variations in the source voltage*. Thus,

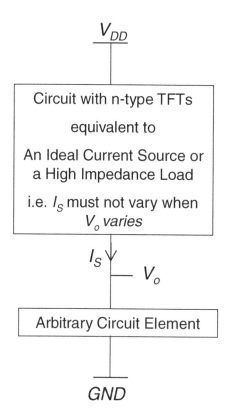

FIGURE 13.5
Illustration of the aim of the design. Due to the absence of a p-type TFT, a circuit substitute for a current source must be designed with n-type TFT. An ideal current source would have high output impedance, i.e. I_S must not vary with varying V_o.

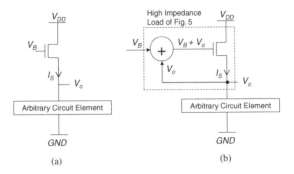

FIGURE 13.6

(a) A single n-type TFT is not an ideal current source, since variations in V_o cause variations in the gate-source voltage which in turn affects I_S. (b) If however, V_o was "added" onto a constant bias V_B at the gate of the n-type TFT, the gate voltage would follow the source voltage. This implies a constant gate-source voltage and hence a constant current, I_S, assuming the TFT is in saturation. (From Sambandan, S.; High-Gain Amplifiers With Amorphous-Silicon Thin-Film Transistors, IEEE Electron Device Letters, IEEE, Volume: 29, Issue: 8 Digital Object Identifier: 10.1109/LED.2008.2000951 Publication Year: 2008, Page(s): 882–884. With permission.)

we need to feedback the output node voltage to the gate of the n-type TFT, such that the gate receives a bias $V_B + V_o$. The current through the circuit will then be

$$I_S = \beta_n((V_B + V_o) - V_o - \upsilon_{Tn})^2 = \beta_n(V_B - \upsilon_{Tn})^2 \qquad (13.2)$$

This current is constant, dependent on V_B, and invariant with changes in V_o.

Now, *how do we ensure the gate voltage to follow the source voltage,V_o?*, i.e. how do we generate the gate voltage $V_B + V_o$? This is achieved by feeding back the source voltage of the n-type TFT though a circuit which *adds* it to V_B and supplies it back to the gate voltage of the TFT as illustrated in Fig. 13.6b. In essence what we desire is a positive feedback of the source voltage (also the output voltage), V_o, to the gate of the TFT. This mechanism together with the n-type TFT is the high impedance circuit of Fig. 13.5.

13.2.1.2   Circuit Design — The "Adder"

We saw that the use of positive feedback as shown in Figure 13.6b can convert an n-type TFT biased in saturation into an ideal current source. *But how does one design a circuit to implement this positive feedback - or "add" - with only n-type TFTs?* This is the aim of this section. There are many ways to construct such a circuit. However, given that we have only n-type TFTs and need to minimise the circuit complexity, we describe one design.

First let us study the familiar circuit shown in Figure 13.7. The circuit is a common source amplifier with n-type TFT load and driver with both TFTs biased in saturation and having the same aspect ratio (and hence the same β_n). The small signal aspects of this circuit was studied earlier in Figure 13.3 where we estimated the dc gain of the amplifier. We now want to study the "large signal" aspects of the circuit - particularly the dependence of V_o on V_B and V_i. Assuming both TFTs have the same threshold voltage, the current through the load TFT is

$$I_{load} = \beta_n(V_B - V_o - v_{Tn})^2 \tag{13.3}$$

and the current through the driver TFT is

$$I_{driver} = \beta_n(V_i - v_{Tn})^2 \tag{13.4}$$

Since the current through the load and driver TFTs must be the same, $I_{load} = I_{driver}$, and

$$\beta_n(V_B - V_o - v_{Tn})^2 = \beta_n(V_i - v_{Tn})^2 \Rightarrow V_o = V_B - V_i \tag{13.5}$$

Note that in order to bias both the TFTs in saturation, we need $V_{DD} \geq V_B - v_{Tn}$ and $V_o \geq V_i - v_{Tn}$. Since $V_o = V_B - V_i$, setting $V_{DD} = V_B > 2V_i$, can be used as a thumb-rule to keep all TFTs in saturation.

Figure 13.7b shows the transfer characteristics of the amplifier with both the load and driver TFT having the same aspect ratio. The bias voltage $V_B = V_{DD}$, and transfer characteristic curves are presented for various $V_{DD} = V_B$ - A:V_B=30V, B:V_B=25V, C:V_B=20V, D:V_B=15V, E:V_B=10V. The transfer curves are compared to the theoretical estimate $V_o = V_B - V_i$ shown in solid lines for the different conditions. This estimate is valid only in the region where both TFTs are biased in saturation. We see that in this region, the circuit does behave close to the theoretical estimate.

Next, let us study the circuit shown in Fig. 13.8a. The circuit is a cascade of two common source amplifiers with n-type load and driver. Once again, all TFTs are biased in saturation, have the same aspect ratio, and the same threshold voltage. The circuit receives three external inputs - V_B, V_i, and V_H, and generates an output V_{o2}. From our analysis of the circuit of Figure 13.7a, we see that

$$V_{o1} = V_B - V_i \tag{13.6}$$

$$V_{o2} = V_H - V_{o1} = (V_H - V_B) + V_i \tag{13.7}$$

Note that the circuit *adds* the inputs, V_i and $V_H - V_B$. This is infact what we require to replace the "adder" block of Figure 13.6. However, since there is some signal loss from the two stages of the amplifier, we would see a fraction, γ, of the theoretical estimate for V_{o2} so that $V_{o2} = \gamma((V_H - V_B) + V_i)$. Ideally, $\gamma = 1$. Figure 13.8 shows the transfer characteristics of the adder block with

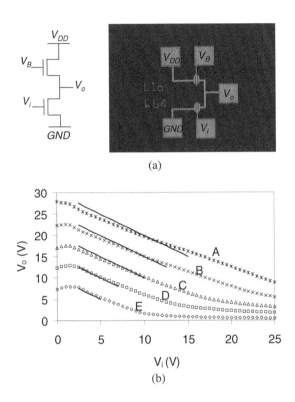

FIGURE 13.7

(a) A common source amplifier circuit and micrograph with n-type TFT load and n-type TFT driver. Both load and driver have an aspect ratio of 64um/16um. The bias $V_B = V_{DD}$. (b) The transfer characteristics of the common source amplifier under different conditions - A:V_B=30V, B:V_B=25V, C:V_B=20V, D:V_B=15V, E:V_B=10V. The solid lines indicate the expected theoretical estimate of $V_o = V_B - V_i$ when both TFTs are in saturation. The experimental observation closely follows theory and thus this circuit is a *subtractor*. (From Sambandan, S.; High-Gain Amplifiers With Amorphous-Silicon Thin-Film Transistors, IEEE Electron Device Letters, IEEE, Volume: 29, Issue: 8 Digital Object Identifier: 10.1109/LED.2008.2000951 Publication Year: 2008, Page(s): 882–884. With permission.)

$V_H = V_{DD} = 30V$ under different bias conditions - A:V_B=15V, B:V_B=20V. The transfer curves are compared to the theoretical estimate $V_{o2} = \gamma((V_H - V_B) + V_i)$ shown in solid lines for the different conditions with $\gamma = 0.7$. The experimental observation closely follows theory.

13.2.1.3 Circuit design — The n-type Current Source

We can now use the adder circuit of Figure 13.8 to replace the conceptual adder block of Figure 13.6b. Hence we can construct a circuit which behaves like an almost ideal current source as shown in Figure 13.9. This circuit effectively behaves like a p-type device sourcing current from the supply to some arbitrary circuit element and in that sense is a *pseudo-complementary device*. If there is variation in V_o, due to variation in the load element, this variation is fed back (i.e added) to the gate of sourcing n-type TFT in order to maintain a constant current.

Figure 13.10 shows the performance of the current source circuit comprised of only n-type TFTs with and without feedback (i.e. feedback disconnected). Without feedback we obtain significant variation in the current as the source voltage is varied. This is essentially the behaviour of the n-type TFT. On the other hand, in the presence of feedback we obtain a region of high impedance where the current through the TFT is invariant with changes in the source voltage, i.e., we have an almost ideal current source.

Let us now consider the small signal analysis of the feedback circuit which is defined by

$$F = \frac{g_{m1}g_{m3}}{(g_{m2} + 2/r_o)(g_{m4} + 2/r_o)} \tag{13.8}$$

The feedback circuit effectively improves the output impedance of the load TFT. The modified output impedance of the load TFT can be shown to be $z_{out} = (r_o - 1 + g_{mL}(1 - F)) - 1$. Without feedback, $F = 0$, the output impedance of the load TFT is $z_{out} = g_{mL} - 1$ as discussed in the case of a conventional common source amplifier. With unity gain feedback, $F = 1$, the output impedance is increased to r_o.

13.2.1.4 Circuit Design — The High Gain Amplifier

If the circuit of Figure 13.9 is used as a load element of a common source amplifier with n-type load, it effectively plays the role of the p-type device since it is a current source. Figure 13.11 illustrates the design of such a common source amplifier. Figure 13.12a shows the transfer characteristics of this amplifier under different bias conditions. The dc gain is extracted from Figure 13.12a and shown in Figure 13.12b. The dc gain of the amplifier is significantly higher than the common source amplifier with n-type load and driver TFTs.

The frequency performance of the high gain amplifier as compared to the common source amplifier with n-type load and driver is limited by the ac characteristics of the feedback circuit. Figure 13.13 compares the step response of the common source amplifier with n-type load and the high gain amplifier under similar loading conditions. It is seen that the high gain amplifier has a time constant about 10 times more than the common source amplifier with n-type load thereby making a tradeoff between speed and dc gain.

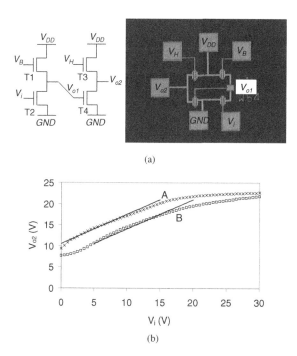

(a)

(b)

FIGURE 13.8

(a) A cascade of two common source amplifiers of Figure 7a. The bias $V_H = V_{DD} = 30V$. (b) The transfer characteristics of the circuit under different conditions - A:V_B=15V, B:V_B=20V. The solid lines indicate the expected theoretical estimate of $V_o = \gamma((V_H - V_B) + V_i)$ when both TFTs are in saturation. The experimental observation closely follows theory and this circuit is an *adder*. (From Sambandan, S.; High-Gain Amplifiers With Amorphous-Silicon Thin-Film Transistors, IEEE Electron Device Letters, IEEE, Volume: 29, Issue: 8 Digital Object Identifier: 10.1109/LED.2008.2000951 Publication Year: 2008, Page(s): 882–884. With permission.)

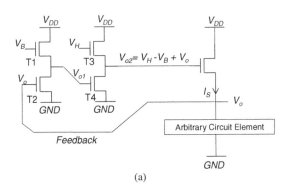

(a)

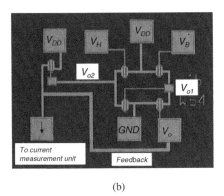

(b)

FIGURE 13.9

(a) The adder circuit of Figure 13.8a is used to replace the conceptual adder in the circuit shown in Figure 13.6b. The adder adds the source (and output) voltage, V_o to a constant bias $V_H - B_B$ and feeds the sum to the gate of the n-type load TFT. This ensures that the gate-source voltage in the n-type TFT is constant at $V_H - V_B$. (b) Micrograph of the circuit. For test, we vary V_o and study the change in current sourced by the n-type load. (From Sambandan, S.; High-Gain Amplifiers With Amorphous-Silicon Thin-Film Transistors, IEEE Electron Device Letters, IEEE, Volume: 29, Issue: 8 Digital Object Identifier: 10.1109/LED.2008.2000951 Publication Year: 2008, Page(s): 882–884. With permission.)

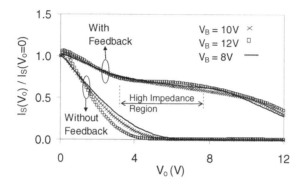

FIGURE 13.10

Test of the current source of Figure 13.1. For test, we vary V_o and study the change in current sourced by the n-type TFT with and without feedback. With feedback, when all TFTs are in saturation, the current does not depend strongly on V_o and the circuit behaves like an ideal current source. On the other hand, when feedback is removed, the current is strongly dependent on V_o. Thus, the presence of positive feedback makes the n-type TFT achieve high output impedance in the region where all TFTs of the feedback circuit are in saturation. (From Sambandan, S.; High-Gain Amplifiers With Amorphous-Silicon Thin-Film Transistors, IEEE Electron Device Letters, IEEE, Volume: 29, Issue: 8 Digital Object Identifier: 10.1109/LED.2008.2000951 Publication Year: 2008, Page(s): 882–884. With permission.)

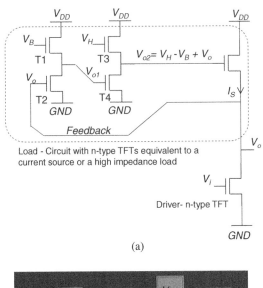

(a)

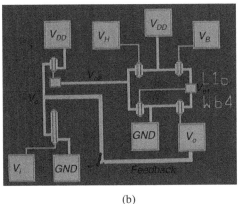

(b)

FIGURE 13.11

(a) The current source of Figure 13.10 can be used as a replacement for a p-type load in a common source amplifier. The presence of this high impedance load is expected to result in high dc gain. (b) Micrograph of the high gain amplifier. All TFTs are 64um/16um except the driver TFT of the amplifier which is 256um/16um. (From Sambandan, S.; High-Gain Amplifiers With Amorphous-Silicon Thin-Film Transistors, IEEE Electron Device Letters, IEEE, Volume: 29, Issue: 8 Digital Object Identifier: 10.1109/LED.2008.2000951 Publication Year: 2008, Page(s): 882–884. With permission.)

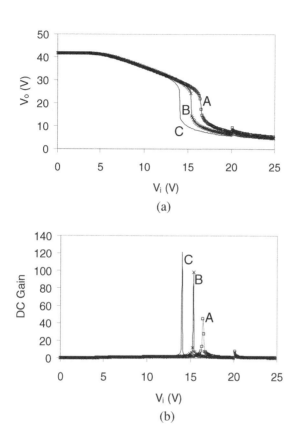

FIGURE 13.12
(a) Transfer characteristics of the high gain amplifier under different bias conditions. $V_H = V_{DD} = 40V$. A: $V_B=12V$, B: $V_B=10V$, C: $V_B=8V$. (b) The dc gain of the amplifier extracted from the transfer characteristics. (From Sambandan, S.; High-Gain Amplifiers With Amorphous-Silicon Thin-Film Transistors, IEEE Electron Device Letters, IEEE, Volume: 29, Issue: 8 Digital Object Identifier: 10.1109/LED.2008.2000951 Publication Year: 2008, Page(s): 882–884. With permission.)

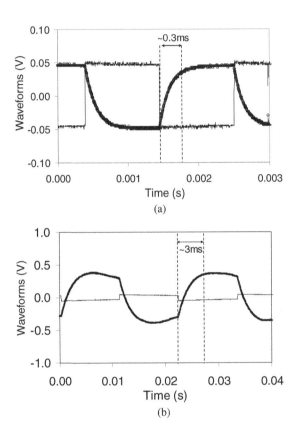

FIGURE 13.13
Pulse response of (a) the common source amplifier of Figure 13.7. (b) the high
gain amplifier of Figure 11. The output of both amplifiers see the same capacitive load. (From Sambandan, S.; High-Gain Amplifiers With Amorphous-
Silicon Thin-Film Transistors, IEEE Electron Device Letters, IEEE, Volume:
29, Issue: 8 ?Digital Object Identifier: 10.1109/LED.2008.2000951 ?Publica-
tion Year: 2008, Page(s): 882–884. With permission.)

Part IV

Appendix

14

Appendix — Derivation of the Threshold Voltage Shift Model

CONTENTS

14.1 State Space Form of Charge Trapping

$$\begin{bmatrix} dn_f/dt \\ dn_s/dt \end{bmatrix} = \begin{bmatrix} -\alpha - \beta & -\alpha \\ -\gamma & -\gamma \end{bmatrix} \begin{bmatrix} n_f \\ n_s \end{bmatrix} + \begin{bmatrix} \alpha \\ \gamma \end{bmatrix} N(t) \qquad (14.1)$$

For a constant gate voltage applied at some time $t = 0$, $N(t)$ assumes the form of a step input such that

$$\begin{aligned} N(t) &= ((V_{gs} - V_{T0})C_i/q)U(t) & (14.2) \\ U(t) &= 0 \ \text{ if } \ t < 0 \\ &= 1 \ \text{ if } \ t \geq 0 \end{aligned}$$

Here, q is the electron charge, C_i is the TFT dielectric capacitance per unit area. Substituting for $N(t)$ and taking the Laplace transform on both sides of Eq. 14.1,

$$\begin{bmatrix} sn_f(s) \\ sn_s(s) \end{bmatrix} - \begin{bmatrix} n_f(0) \\ n_s(0) \end{bmatrix} = \begin{bmatrix} -\alpha - \beta & -\alpha \\ -\gamma & -\gamma \end{bmatrix} \begin{bmatrix} n_f(s) \\ n_s(s) \end{bmatrix} + \begin{bmatrix} \alpha \\ \gamma \end{bmatrix} N(s) \ (14.3)$$

Assuming there are no trapped charges at $t = 0$, $n_f(0) = n_s(0) = 0$, and Eq. 14.3

$$\begin{bmatrix} n_f(s) \\ n_s(s) \end{bmatrix} = \begin{bmatrix} s + \alpha + \beta & \alpha \\ \gamma & s + \gamma \end{bmatrix}^{-1} \begin{bmatrix} \alpha \\ \gamma \end{bmatrix} N(s) \qquad (14.4)$$

14.2   Solving for $n_f(t)$ and $n_s(t)$

We can define the state transition matrix $\mathbf{\Phi}$ as

$$\mathbf{\Phi}(t) \;=\; L^{-1}\left(\begin{bmatrix} s+\alpha+\beta & \alpha \\ \gamma & s+\gamma \end{bmatrix}^{-1}\right) \tag{14.5}$$

$$=\; L^{-1}\left(\frac{1}{(s+\lambda_1)(s+\lambda_2)}\begin{bmatrix} s+\gamma & -\alpha \\ -\gamma & s+\alpha+\beta \end{bmatrix}\right)$$

where $2\lambda_1 = \alpha + \beta + \gamma - ((\alpha+\beta+\gamma)^2 - 4\beta\gamma)^{1/2}$ and $2\lambda_2 = \alpha + \beta + \gamma + ((\alpha+\beta+\gamma)^2 - 4\beta\gamma)^{1/2}$. Note that $\lambda_2 > \lambda_1 > 0$. Taking the inverse Laplace transform, we find the state transition matrix to be

$$\mathbf{\Phi}(t) = \frac{1}{\lambda_1 - \lambda_1}$$
$$\times \begin{bmatrix} -(\lambda_1-\gamma)e^{-\lambda_1 t} + (\lambda_2-\gamma)e^{-\lambda_2 t} & -\alpha e^{-\lambda_1 t} + \alpha e^{-\lambda_2 t} \\ -\gamma e^{-\lambda_1 t} + \gamma e^{-\lambda_2 t} & -(\lambda_1-\alpha-\beta)e^{-\lambda_1 t} + (\lambda_2-\alpha-\beta)e^{-\lambda_2 t} \end{bmatrix} \tag{14.6}$$

The solution to Eq. 14.1 is then given by

$$\begin{bmatrix} n_f(t) \\ n_s(t) \end{bmatrix} = \int_0^t \mathbf{\Phi}(t-\tau)\begin{bmatrix} \alpha \\ \gamma \end{bmatrix}\frac{(V_{gs}-V_{T0})C_i}{q}d\tau \tag{14.7}$$

Solving this equation, we find

$$n_f(t) \;=\; \frac{(V_{gs}-V_{T0})C_i}{q}\frac{1}{\lambda_2-\lambda_1}\left[-\alpha(1-e^{-\lambda_1 t}) + \alpha(1-e^{-\lambda_2 t})\right] \tag{14.8}$$

$$n_s(t) \;=\; \frac{(V_{gs}-V_{T0})C_i}{q}\frac{1}{\lambda_2-\lambda_1}$$
$$\times \left[\left(-\gamma + \frac{\beta\gamma}{\lambda_1}\right)(1-e^{-\lambda_1 t}) + \left(\gamma - \frac{\beta\gamma}{\lambda_2}\right)(1-e^{-\lambda_2 t})\right]$$

14.3   Threshold Voltage Shift

The threshold voltage shift is related to the total trapped charge as

$$\delta V_T(t) = \frac{q(n_f(t)+n_s(t))}{C_i} \tag{14.9}$$

Noting that $\beta\gamma = \lambda_1\lambda_2$ we obtain from Eq. 14.8,

$$\delta V_T(t) = (V_{gs}-V_{T0})(1 - \varphi e^{-\lambda_1 t} - (1-\varphi)e^{-\lambda_2 t}) \tag{14.10}$$

where $\varphi = \frac{\lambda_2-\alpha-\gamma}{\lambda_2-\lambda_1}$. Since $\alpha > 0$, $\beta > 0$ and $\gamma > 0$, it can be easily shown that $0 < \varphi < 1$.

14.4 Generalizations

Note that the general linear for the model for threshold voltage shift as defined by Eq. 14.1 can be modified to include trapping in multiple states.

However, the most powerful feature of this model is that it can be used to predict the threshold voltage shift for almost any gate bias. For example, we have considered the case for a constant gate bias. If however, the gate bias has a sinusoidal behavior with time, or a linear ramp line behavior with time, etc., the above model is still valid, with $N(t)$ taking different functions for each input.

This model can therefore provide a powerful circuit simulator for the threshold voltage shift in TFTs.

Bibliography

[1] L. Chua, *IEEE Trans. Circuit Theory*, Memristor—the missing circuit element, 18:507–519, 1971.

[2] D. B. Strukov, G. S. Snider1, D. R. Stewart and R. S. Williams, *Nature*, "Memristive" switches enable "stateful" logic operations via material implication, 453:80–83, 2007.

[3] S. M. Sze, *Physics of Semiconductor Devices*, John Wiley and Sons Inc, 2002.

[4] D. K. Schroder, *Semiconductor Material and Device Characterization*, John Wiley and Sons Inc, 2006.

[5] B. Razavi, *Design of Analog CMOS Integrated Circuits*, McGraw-Hill, 2001.

[6] P. R. Gray, P. J. Hurst, S. H. Lewis, and R. G. Meyer, *Analysis and Design of Analog Integrated Circuits*, John Wiley and Sons, Inc, 2001.

[7] J. M. Rabaey, A. Chandrakasan, and B. Nikolic, *Digital Integrated Circuits*, John Wiley and Sons, Inc, 2003.

[8] R. Zallen, *The Physics of Amorphous Solids*, John Wiley and Sons Inc, 1983.

[9] M. H. Brodsky, *Amorphous Semiconductors*, Springer-Verlag, Berlin, 1979.

[10] G. A. N. Connell, and G. Lucovsky, *J. Non-Crystalline Solids*, Structural models for amorphous semiconductors and insulators, 31:123, 1978.

[11] N. F. Mott, and E. A. Davis, *Electronic Processes in Noncrystalline Materials*, Clarendon Press, Oxford, 1979.

[12] R. A. Street, *Hydrogenated Amorphous Silicon*, Cambridge University Press, 1991.

[13] T. Tiedje and A. Rose, *Solid State Communications*, A physical interpretation of dispersive transport in disordered semiconductors, 37:49–52, 1981.

[14] P. G. Le Comber and W.E. Spear, *Phys. Rev. Lett.*, Electronic transport in amorphous silicon films, 25:509–511, 1970.

[15] W.E. Spear andP.G. Le Comber, *Journal of Non-Crystalline Solids*, Investigation of the localised state distribution in amorphous Si films, 8–10:727–7380, 1972.

[16] C.Y. Huang, S. Guha and S.J. Hubgens, *Physical Review*, Study of gap states in hydrogenated amorphous silicon by transient and study state photoconductivity measurements, 27:7460–7465, 1983.

[17] J.D. Cohen, D.V. Lang and J.P. Harbison, *Phys. Rev. Lett*, Direct measurement of the bulk density of gap states in n-type hydrogenated amorphous silicon, 45:197–200, 1980.

[18] M. Hirose, T. Suzuki and G.H. Dohler, *Applied Physics Letters*, Electronic density of states in discharge produced amorphous silicon, 34:234–236, 1979

[19] J.G. Shaw and M. Hack, *Journal of Applied Physics*, An analytical theory model for calculating trapped charge in amorphous silicon, 64:4562–4566, 1988.

[20] M.J. Powell, *Philosophical Magazine*, Analysis of field effect conductance measurements on amorphous semiconductors, 43:93–103, 1981.

[21] P. Viktorovitch and G. Moddel, *Journal of Applied Physics*, Interpretation of the conductance and capacitance frequency dependence of hydrogenated amorphous silicon Schottky barrier diodes, 51:4847–4854, 1980.

[22] M. Shur and M. Hack, *Journal of Applied Physics*, Physics of amorphous silicon based alloy field effect transistors, 55:3831–3842, 1984.

[23] S. Kishida, Y. Naruke, Y. Uchida and M. Matsumura, *Journal of Applied Physics*, Theoretical analysis of amorphous silicon field effect transistors, 22:511–517, 1983.

[24] M.S. Shur, M.D. Jacunski, H.C. Slade and M. Hack, *Journal of the Society for Information*, Analytical models for amorphous-silicon and polysiliconthin-film transistors for high definition display technology, 1:223–236, 1995.

[25] P. Servathi, D. Striakhilev and A. Nathan, *Electron Devices, IEEE*, Above threshold modeling and parameter extraction for amorphous silicon thin film transistors, 50:727–738, 2003.

[26] M.S. Shur, H.C. Slade, M.D. Jacunski, A.A. Owusu and T. Ytterdal, *Journal of the Electrochemical Society*, SPICE models for amorphous silicon and poly silicon thin film transistors, 144:2833–2839, 1997.

[27] M. Shur, M. Hack and J.G. Shaw, *Journal of Applied Physics*, A new analytical model for amorphous silicon thin film transistors, 66:3371–3380, 1989.

[28] T. Leroux, *Solid-state Electronics*, Static and dynamic analysis of amorphous silicon field effect transistors, 29:47–58, 1986.

[29] K. Khakzar and E.H. Lueder, *Electron Devices, IEEE Transactions*, Modeling of amorphous-silicon thin film transistors for circuit simulations with SPICE, 39:1428–1434, 1992.

[30] S.S. Chen and J.B. Kuo, *Electron Devices, IEEE Transactions*, An analytical a Si:H TFT DC/Capacitance model using an effective temperature approach for deriving a switching time model for an inverter circuit considering deep and tail states, 41:1169–1178, 1994.

[31] K. Khakzar, *Electron Devices, IEEE Transactions*, Modeling of amorphous-silicon thin-film transistors for circuit simulations with SPICE, 39:1428–1434, 1992.

[32] B. Iniguez, T.A. Fjeldly and M.S. Shur, *Solid-State Electronics*, Thin film transistor modeling, 43:703–723, 1999.

[33] L. Colalongo, *Solid-State Electronics*, A new analytical model for amorphous silicon thin film transistors including tail and deep states, 45:1525–1530, 2001.

[34] P. Servati, D. Striakhilev and A. Nathan, *Electron Devices, IEEE Transactions*, Above threshold parameter extraction including contact resistance effects for Si:H TFTs on glass and plastic, 39:762, 2003.

[35] H.C. Slade, M.S. Shur, S.C. Deane and M. Hack, *MRS Proceedings*, Physics of below threshold current distribution in Si:H TFTs, 420:257–262, 1996.

[36] H.C. Slade, M.S. Shur, S.C. Deane and M. Hack, *MRS Proceedings*, Below threshold conduction in a Si:H thin film transistors with and without a silicon nitride passivating layer, 420:2560–2562, 1996.

[37] P. Servati and A. Nathan, *Electron Devices*, Modeling of the reverse characteristics of a Si:H TFTs, 49:812–819, 2002.

[38] T. Holstein, *Ann. Phys.*, Studies of polaron motion: Part II. The "small" polaron, 8:343–389, 1959.

[39] D.F. Barbe and C.R. Westgate, *J. Phys. Chem. Solid*, Surface state parameters of metal-free phthalocyanine single crystals, 31:2679–2687, 1970.

[40] M. L. Petrova and L. D. Rozenshtein, *Sov. Phys. Solid State*, Field effect and slow states in thin films of organic semiconductors, 1972.

[41] F. Ebisawa, T. Kurokawa and S. Nara, *J. Appl. Phys*, Electrical properties of polyacetylene/polysiloxane interface, 54:3255–3259, 1983.

[42] H. Koezuka, A. Tsumura, and T. Ando, *Synth. Met*, Field-effect transistor with polythiophene thin film, 25:699–704, 1987.

[43] D. D. Eley, *Nature*, Phthalocyanines as semiconductors, 162:819, 1948.

[44] F. Garnier, X. Peng, G. Horowitz, and D. Fichou, *Molecular Engineering*, Organic-based field-effect transistors: Critical analysis of the semiconducting characteristics of organic materials, 1:131–139, 1991.

[45] H. Shirakawa, E. J. Louis, A. G. MacDiarmid, C. K. Chiang, A. J. Heeger, *J. Chem. Soc. Chem. Com*, Synthesis of electrically conducting organic polymers: Halogen derivatives of polyacetylene(CH) x, 1977, 16:578, 1977.

[46] D. D. Eley, G. D. Parfitt, M. J. Perry, and D. H. Taysum, *Trans. Faraday Soc*, The semiconductivity of organic substances, 49:79–86, 1953.

[47] H. E. Katz, *J. Mater. Chem*, Printable organic and polymeric semiconducting materials and devices, 7:359–369, 1997.

[48] W. R. Salaneck, S. Stafstrom and J. L. Bredas, *Phil. Trans. Roy. Soc. A.*, Conjugated polymer surfaces and interfaces, 355:789–799, 1996.

[49] T. Holstein, *Ann. Phys*, Studies of polaron motion: Part I. The molecular-crystal model, 8:325–342, 1959.

[50] T. Holstein, *Ann. Phys*, Studies of polaron motion: Part II. The "small" polaron, 8:343–389, 1959.

[51] D. M. Pai, J. F. Yanus, M. Stolka, *J. Phys. Chem*, Trap-controlled hopping transport, 88:4714–4717, 1984.

[52] J. Frenkel, *Phys. Rev.*, On pre-breakdown phenomena in insulators and electronic semiconductors, 54:647–648, 1938.

[53] G. Horowitz, X.-Z. Peng, D. Fichou, F. Garnier, *Synth. Met*, Role of the semiconductor/insulator interface in the characteristics of -conjugated-oligomer-based thin-film transistors, 51:419–424, 1992.

[54] C. P. R. Dockendorf, D. Poulikakos, G. Hwang, B. J. Nelson, and C. P. Grigoropoulos, *Phys. Rev. Lett.*, Maskless writing of a flexible nanoscale transistor with Au-contacted carbon nanotube electrodes, 91:243–118, 2007.

[55] Z. Xie, M. S. A. Abdou, X. Lu, M. J. Deen, and S. Holdcroft, *Can. J. Phys*, Electrical characteristics and photolytic tuning of poly (3-hexylthiophene) thin film metal-insulator-semiconductor field-effect transistors (MISFETs), 70:1171, 1992.

[56] W. S. Wong, E. M. Chow, R. Lujan, V. Geluz-Aguilar, and M. L. Chabinyc, *Applied Physics Letters*, Fine-feature patterning of self-aligned polymeric thin-film transistors fabricated by digital lithography and electroplating, 89:142118, 2006.

[57] R. A. Street, W. S. Wong, S. E. Ready, M. L. Chabinyc, A. C. Arias, S. Limb, A. Salleo, and R. Lujan, *Materials Today*, Jet printing flexible displays, 9:32–37, 2006.

[58] L. L. Lavery, G. L. Whiting, and A. C. Arias, *Organic Electronics*, All ink-jet printed polyfluorene photosensor for high illuminance detection, 12:682–685, 2011.

[59] G. L. Whiting, and A. C. Arias, *Applied Physics Letters*, Chemically modified ink-jet printed silver electrodes for organic field-effect transistors, 95:253–302, 2009.

[60] A. C. Arias, J. D. MacKenzie, I. McCulloch, J. Rivnay, and A. Salleo, *Chem. Rev.*, Materials and applications for large area electronics: Solution-based approaches, 110:3–24, 2010.

[61] H. Fuchigami, A. Tsumara, and H. Koezuka, *Appl. Phys. Lett*, Polythienylenevinylene thin film transistor with high carrier mobility, 63:1372, 1993.

[62] J. Paloheimo, P. Kuivalainen, H. Stubb, E. Vuorimaa, and P. Yli-Lahti, *Appl. Phys. Lett*, Molecular field–effect transistors using conducting polymer Langmuir–Blodgett films, 56:1157, 1990.

[63] F. Garnier, R. Hajlaoui, A. Yassar, and P. Srivastava, *Science*, All-polymer field-effect transistor realized by printing techniques, 265:1684, 1994.

[64] J. A. Rogers, Z. Bao, and V. R. Raju, *Appl. Phys. Lett*, Nonphotolithographic fabrication of organic transistors with micron feature sizes, 1998, 72:2716, 1998.

[65] Z. Bao, A. Dodabalapur, and A. J. Lovinger, *Appl. Phys. Lett*, Soluble and processable regioregular poly(3hexylthiophene) for thin film field effect transistor applications with high mobility, 69:4108, 1996.

[66] Z. Bao, Y. Feng, A. Dodabalapur, V. R. Raju, and J. Lovinger, *Chem. Matter*, Performance of poly(3-hexylthiophene) organic field-effect transistors on cross-linked poly(4-vinyl phenol) dielectric layer and solvent effects, 92:1299, 1977.

[67] M. Willander, A. Assadi, and C. Svensson, *Synth. Met*, Polymer based devices their function and characterization, 57:4099–4104, 1993.

[68] J. H. Burroughes, C. A. Jones, and R. H. Friend, *Nature*, New semiconductor device physics in polymer diodes and transistors, 335:137–141, 1988.

[69] J. Paloheimo, H. Stubb, P. Yli-Lahti, P. Kuivalainen, *Synth. Met*, Field-effect conduction in polyalkylthiophenes, 41:563–566, 1991.

[70] L. Brgi, T. J. Richards, R. H. Friend, and H. Sirringhaus, *J. Appl. Phys*, Close look at charge carrier injection in polymer field-effect transistors 94:6129, 2003.

[71] L. Brgi, R. H. Friend, and H. Sirringhaus, *Appl. Phys. Lett*, Formation of the accumulation layer in polymer field-effect transistors, 82:1482, 2003.

[72] G. Horowitz, and P. Delannoy, *J. Appl. Phys*, An analytical model for organic–based thin–film transistors, 70:469, 1991.

[73] A. R. Brown, C. P. Jarrett, D. M. de Leeuw, and M. Matters, *Synth. Met*, Field-effect transistors made from solution-processed organic semiconductors, 88:37, 1997.

[74] K. Waragai, H. Akimichi, S. Hotta, H. Kano, and H. Sakaki, *Synth. Met*, FET characteristics of substituted oligothiophenes with a series of polymerization degrees, 57:4053, 1993.

[75] A. Dodabalapur, L. Torsi and H.E. Katz, *Science*, Organic transistors: Two-dimensional transport and improved electrical characteristics, 268:270–271, 1995.

[76] L. Torsi, A. Dodabalapur, L.J. Rothberg, A.W.P. Fung and H.E. Katz, *Science*, Intrinsic transport properties and performance limits of organic field-effect transistors, 272:1462–1464, 1996.

[77] L. Torsi, A. Dodabalapur and H.E. Katz, *J. Appl. Phys*, An analytical model for short–channel organic thin–film transistors, 78:1088–1093, 1995.

[78] G. Horowitz, Organic field effect transistors, *Adv. Mater*, 10:365–377, 1998.

[79] A. P. Marchetti, K. E. Sassin, R. H. Young, L. J. Rothberg, and D. Y. Kondakov, *J. Appl. Phys*, Integer charge transfer states in organic light-emitting diodes: Optical detection of hole carriers at the anode—organic interface, 109:013709–013717, 2011.

[80] Chabinyc, M. L., R. A. Street, and J. E. Northrup, *Appl. Phys. Lett*, Effects of molecular oxygen and ozone on polythiophene-based thin-film transistors, 90:123508, 2007.

[81] M. L. Chabinyc, R. Lujan, F. Endicott, M. F. Toney, I. McCulloch, and M. Heeney, *Appl. Phys. Lett*, Effects of the surface roughness of plastic-compatible inorganic dielectrics on polymeric thin film transistors, 90:233508, 2007.

[82] M. L. Chabinyc, L. H. Jimison, J. Rivnay and A. Salleo, *M.R.S Bulletin*, Connecting electrical and molecular properties of semiconducting polymers for thin-film transistors, 33:683–689, 2008.

[83] R. A. Street, M. L. Chabinyc, F. Endicott and, B. Ong, *Applied Physics*, Extended time bias stress effects in polymer transistors, 100:114518, 2006.

[84] G. Horowitz, D. Fichou, X. Z. Peng, Z. G. Xu, and F. Garnier, *Solid State Commun*, 72:381, 1989.

[85] H. Akimichi, K. Waragai, S. Hotta, H. Kano, and H. Sakati, *Appl. Phys. Lett*, 58:1500, 1991.

[86] F. Garnier, A. Yassar, R. Hajlaoui, G. Horowitz, F. Deloffre, B. Servet, S. Ries, and P. Alnot, *J. Am. Chem. Soc*, Molecular engineering of organic semiconductors: Design of self-assembly properties in conjugated thiophene oligomers, 115:8716, 1993.

[87] R. Hajlaoui, D. Fichou, G. Horowitz, B. Nessakh, M. Constant, and F. Garnier, *Adv. Mater*, Organic transistors using octithiophene and, dihexyloctithiophene: Influence of oligomer length versus molecular ordering on mobility, 9:557, 1997.

[88] G. Horowitz, B. Bachet, A. Yassar, P. Lang, F. Demanze, J. L. Fave, and F. Garnier, *Chem. Mater*, Growth and characterization of sexithiophene single crystals, 7:1337, 1995.

[89] T. Siegrist, C. Kloc, R. A. Laudise, H. E. Katz, and R. C. Haddon, *Adv. Mater*, Crystal growth, structure, and electronic band structure of 4T polymorphs, 10:379, 1998.

[90] K. Pichler, C. P. Jarrett, R. H. Friend, B. Ratier, and A. Moliton, *J. Appl. Phys*, Field–effect transistors based on poly (p–phenylene vinylene) doped by ion implantation, 77:3523, 1995.

[91] A. R. Brown, D. M. Deleeuw, E. E. Havinga, and A. Pomp, *Synth. Met*, A universal relation between conductivity and field-effect mobility in doped amorphous organic semiconductors, 68:65, 1994.

[92] A. Facchetti, H. Klauk (Editor), *N-Channel Organic Semiconductors for Transistors and Complementary Circuits*, John Wiley and Sons, Inc

[93] H. Yan, Z. Chen, Y. Zheng, C. E. Newman, J. Quin, F. Dolz, M. Kastler, and A. Facchetti, *Nature*, A high-mobility electron-transporting polymer for printed transistors, 457:679–686, 2009.

[94] Y.-Y. Lin, A. Dodabalapur, R. Sarpeshkar, Z. Bao, W. Li, K. Baldwin, V. R. Raju, and H. E. Katz, *Appl. Phys. Lett.*, Organic complementary ring oscillators, 74:2714, 1999.

[95] A. Nathan, A. Kumar, K. Sakariya, P. Servati, S. Sambandan, and D. Striakhilev, *IEEE J. Solid-State Circuits*, Amorphous silicon thin film transistor circuit integration for organic LED displays on glass and plastic, 39:1477–1486, 2004.

[96] T. N. Ng, W. S. Wong, M. L. Chabinyc, S. Sambandan and R. A. Street, *Appl. Phys. Lett*, Flexible image sensor array with bulk heterojunction organic photodiode, 92:213303, 2008.

[97] M.D. Angione, S. Cotrone, M. Magliulo, A. Mallardi, D. Altamura, C. Giannini, N. Cioffi, L. Sabbatini, E. Fratini, P. Baglioni, G. Scamarcio, G. Palazzo and L. Torsi *Proc. Nat. Acad. Sci*, Interfacial electronic effects in functional bio-layers integrated into organic field-effect transistors, 109:6429–6434, 2012

[98] C. D. Dimitrakopoulos, S. Purushothaman, J. Kymissis, A. Callegari and J. M. Shaw, *Science*, Low-voltage organic transistors on plastic comprising high-dielectric constant gate insulators, 283:822–824, 1999.

[99] D.-H. Kim, J.-H. Ahn, W.-M. Choi, H.-S. Kim, T.-H. Kim, J. Song, Y.Y. Huang, L. Zhuangjian, L. Chun, and J.A. Rogers, *Science*, Stretchable and foldable silicon integrated circuits, 320:507–511, 2008.

[100] H. Gleskova, S. Wagner, W. Soboyejo, and Z. Suo, *J. Appl. Phys*, Electrical response of amorphous silicon thin-film transistors under mechanical strain, 92:6224–6229, 2002.

[101] H. Gleskova and S. Wagner, *Applied Physics Lett*, Electron mobility in amorphous silicon thin-film transistors under compressive strain, 79:3347–3349, 2001.

[102] P.I. Hsu, H. Gleskova, M. Huang, Z. Suo, S. Wagner, and J.C. Sturm, *J. Non-Cryst. Solids*, Amorphous Si TFTs on plastically-deformed spherical domes, 299–302:1355–1359, 2002.

[103] P. I. Hsu, R. Bhattacharya, H. Gleskova, M. Huang, Z. Xi, Z. Suo, S. Wagner, and J. C. Sturm, *Applied Physics Lett*, Thin-film transistor circuits on large-area spherical surfaces, 81:1723–1725, 2002.

[104] A. Madan, P. G. Le Comber, and W. E. Spear, *J. Non Cryst. Solids*, Investigation of the density of states in a-Si using the field effect technique, 20:239, 1976.

[105] C. Longeaud, J. A. Schmidt, R. R. Koropecki, and J. P. Kleider, Determination of the hydrogenated amorphous silicon density of states parameters from photoconductivity measurements, *J. Optoelectronics Adv. Mat.*, 11:1064–1071, 2009.

[106] D.V. Lang, Deep-level transient spectroscopy: A new method to characterize traps in semiconductors, *J. Appl. Phys.*, 45:3023, 1974.

[107] M.J. Powell, *IEEE Trans. Electron. Devices*, The physics of amorphous-silicon thin-film transistors, 36:2753, 1989.

[108] C. van Berkel, M.J. Powell, *Appl. Phys. Lett.*, 51:1094, 1987.

[109] W.B. Jackson and M.D. Moyer, *Phys. Rev.B*, Creation of near-interface defects in hydrogenated amorphous silicon-silicon nitride heterojunctions: The role of hydrogen, 36:6217–6220, 1987.

[110] S.C. Deane, R. B. Wehrspohn and M.J. Powell, *Phys. Rev.B*, Unification of the time and temperature dependence of dangling-bond-defect creation and removal in amorphous-silicon thin-film transistors, 58:12625–12628, 1998.

[111] M. Stutzmann, W. B. Jackson, and C. C. Tsai, *Phys. Rev. B*, Light-induced metastable defects in hydrogenated amorphous silicon: A systematic study, 32:23–47, 1985.

[112] Y. Kaneko, A. Sasano and T. Tsukada, *J. Appl. Phys*, Characterization of instability in amorphous silicon thin–film transistors, 69:7301–7305, 1991.

[113] R.B. Wehrpohn, S.C. Deane, I.D. French, I.G. Gale, R. Bruggemann and M.J. Powell, Urbach energy dependence of the stability in amorphous silicon thin-film transistors, *Appl. Phys. Lett*, 74:3374–3376, 1999.

[114] R.B. Wehrpohn, S.C. Deane, I.D. French, I.G. Gale, M.J. Powell and J. Robertson, *J. Appl. Physics*, Relative importance of the Si–Si bond and Si–H bond for the stability of amorphous silicon thin film transistors, 87, 2000.

[115] R.B. Wehrpohn, S.C. Deane, I.D. French, I.G. Gale and M.J. Powell, *J. Non-Cryst. Solids*, Effect of amorphous silicon material properties on the stability of thin film transistors: evidence for a local defect creation model, 266–269:459–463, 2000.

[116] K.S. Karim, A. Nathan, M. Hack and W.I. Milne, *IEEE Electron Device Letts,* Drain bias dependence of threshold voltage stability of amorphous silicon TFTs, 25-:188–190, 2004.

[117] H.L. Gomes, P. Stalinga, F. Dinelli, M. Murgia, F. Biscarini, D.M. de Leeuw, T. Much, J. Geurb, L.W. Molenkamp and V. Wagner, Bias-induced threshold voltages shifts in thin-film organic transistors , *Appl. Physics. Lett,* 84, 2004.

[118] K. D. Mackenzie, P. G. Le Comber and W. E. Spear, *Philos. Mag. Part. B,* The density of states in amorphous silicon determined by space-charge-limited current measurements, 46:377–389, 1982.

[119] A. Madan, P.G. Le Comber, and W.E. Spear, *J. Non-Cryst. Solids,* Investigation of the density of localized states in a-Si using the field effect technique, 20:239–257, 1976.

[120] M. Kuhn, *Solid-State Electronics,* A quasi-static technique for MOS C-V and surface state measurements, 13:873–88, 1970.

[121] R. A. Abram, and P. J. Doherty, *Philos. Mag. Part B,* A theory of capacitance-voltage measurements on amorphous silicon Schottky bar-riers, 45:167–176, 1982.

[122] J. S. Choi, *IEEE Trans. Electron Devices,* Frequency-dependent capacitance-voltage characteristics for amorphous silicon-based metal-insulator-semiconductor structures 39:2515–2522, 1992.

[123] C.-Y. Huang, S. Guha, and S. J. Hudgens, *Phys. Rev. B,* Study of gap states in hydrogenated amorphous silicon by transient and steady-state photoconductivity measurements 27:7460–7465, 1983.

[124] D.V. Lang, *J. Appl. Phys,* Deep-level transient spectroscopy: A new method to characterize traps in semiconductors, 45:3023, 1974.

[125] D. V. Lang, J. D. Cohen, and J. P. Harbison, *Phys. Rev. B,* Measure-ment of the density of gap states in hydrogenated amorphous silicon by space charge spectroscopy, 25:5285–5320, 1982.

[126] J. D. Cohen, D. V. Lang, and J. P. Harbison, *Phys. Rev. Lett,* Direct measurement of the bulk density of gap states in n-type hydrogenated amorphous silicon, 45:197–200, 1980.

[127] S. Sambandan, L. Zhu, D. Striakhilev, P. Servati, and A. Nathan, *IEEE Electron Device Lett,* Markov model for threshold-voltage shift in amor-phous silicon TFTs for variable gate bias, 26:375–377, 2005.

[128] S. Sambandan, and A. Nathan, *IEEE Trans. Electron Device,* Equiv-alent circuit description of threshold voltage shift in a-Si:H TFTs from a probabilistic analysis of carrier population dynamics, 53:2306–2311, 2006.

[129] G. Wegmann, E.A. Vittoz, and F. Rahali, *IEEE J. Solid State Circuits*, Charge injection in analog MOS switches, 22:1091–1097, 1987.

[130] A. Kumar, S. Sambandan, K. Sakariya, P. Servati, A. Nathan, *J. Vacuum Science & Technology A: Vacuum, Surfaces, and Films*, Amorphous silicon shift registers for display drivers, 22:981–986, 2004.

[131] L. T. Clark, B. D. Vogt, R. Shringarpure, S. M. Venugopal, S. G. Uppili, K. Kaftanoglu, H. Shivalingaiah, Z. P. Li, R. J. J. Fernando, E. J. Bawolek, S. M. O'Rourke, *IEEE Trans. Electron Devices*, Circuit-level impact of a-Si:H thin-film-transistor degradation effects, 56:1166–1176, 2009.

[132] M. A. Marrs, C. D. Moyer, E. J. Bawolek, R. J. Cordova, J. Trujillo, G. B. Raupp, B. D. Vogt, *IEEE Trans. Electron Devices*, Control of threshold voltage and saturation mobility using dual-active-layer device based on amorphous mixed metal–oxide–semiconductor on flexible plastic substrates, 58:3428–3434, 2011.

[133] Y.-C. Tarn, P.-C. Ku, H.-H. Hsieh, and L.-H. Lu, *IEEE J. Solid State Circuits*, An amorphous-silicon operational amplifier and its application to a 4-bit digital-to-analog converter, 45:1028–1035, 2010.

[134] C.-H. Wu, H.-H. Hsieh, P.-C. Ku, and L.-H. Lu, *IEEE J. Display Technology*, A differential Sallen-Key low-pass filter in amorphous-silicon technology, 6:207–214, 2010.

[135] S.M. Venugopal, and D.R. Allee, *IEEE J. Display Technology*, Integrated a-Si:H source drivers for 4"QVGA electrophoretic display on flexible stainless steel substrate, 3:57–63, 2007.

[136] C.-H. Kim, S.-J. Yoo, H.-J. Kim, J.-M. Jun, and J.-Y. Lee, *J. Soc. Inf. Display*, Integrated a-Si TFT row driver circuits for high-resolution applications, 14:333–337, 2006.

[137] S. Sambandan, and A. Nathan, *IEEE Proc. Solid State Circuit Conference*, Circuit techniques for organic and amorphous semiconductor based field effect transistors, 69–72, 2006.

[138] S. Sambandan, A. Kumar, K. Sakariya, and A. Nathan, *IEE Electronic Lett*, Analogue circuit building blocks with amorphous silicon thin film transistors, 41:314–315, 2005.

[139] S.Y. Yoon, Y.H. Jang, B. Kim, M.D. Chun, H.N. Cho, N.W. Cho, C. Y. Sohn, S.H. Jo, C.-D. Kim, and I.-J. Chung, *Soc. Inf. Display Dig*, Highly stable integrated gate driver circuit using a-Si TFT with dual pull-down structure, 36:348–351, 2005.

[140] T. Ng, S. Sambandan, R. Lujan, A. Arias, C. R. Newman, H. Yan, A. Facchetti, *Appl. Phys. Lett*, Electrical stability of inkjet-patterned organic complementary inverters measured in ambient conditions, 94:233–307, 2009.

[141] S. Sambandan, *IEEE Electron Device Lett*, High-gain amplifiers with amorphous-silicon thin-film transistors, 29:882–884, 2010.

[142] H. Lebrun, T. Kretz, J. Magarino, and N. Szydlo, *Soc. Inf. Display Dig*, Design of integrated drivers with a-Si TFTs for small displays: Basic concepts, 36:950–953, 2005.

[143] B.-S. Bae, J.-W. Choi, J.-H. Oh, and J. Jang, *IEEE Trans. Electron Devices*, Level shifter embedded in drive circuits with amorphous silicon TFTs, 53:494–498, 2006.

[144] V. M. Da Costa and R. A. Martin, *IEEE J. Solid-State Circuits*, Amorphous silicon shift register for addressing output drivers, 29:596–600, 1994.

[145] P. Madeira and R. Hornsey, *IEEE CCECE*, Analog circuit design using amorphous silicon thin film transistors, 2:633–636, 1997.

[146] R.A. Street, *Technology and Application of Amorphous Silicon*, Springer, 1999.

[147] M. Stewart, P. A. Allentown, R. S. Howell, L. Piers, and M. K. Hatalis, *IEEE Trans. Electron Devices*, Polysilicon TFT technology for active matrix OLED displays, 48:845–851, 2001.

[148] W. Wong, T.N. Ng, S. Sambandan, M. Chabinyc, *IEEE Design and Test of Computers*, Materials, processing, and testing of flexible image sensor arrays, 28:16–23, 2011.

[149] T. Ng, R. Lujan, S. Sambandan, R. A. Street, S. Limb, W. Wong, *Appl. Phys. Letts*, Low temperature a-Si:H photodiodes and flexible image sensor arrays patterned by digital lithography, 91:063505, 2007.

[150] T. Someya, Y. Kato, S. Iba, T. Sekitani, Y. Noguchi, H. Kawaguchi, and T. Sakurai, *Electron Devices, IEEE Transactions*, Integration of organic FETs with organic photodiodes for a large area, flexible, and lightweight sheet image scanners, 52:2502–2511, 2005.

[151] T. Someya, T. Sekitani, S. Iba, Y. Kato, H. Kawaguchi, and T. Sakurai, *PNAS*, A large-area, flexible pressure sensor matrix with organic field-effect transistors for artificial skin applications, 101:9966–9970, 2004.

[152] T. Someya, H. Kawaguchi, and T. Sakurai, *IEEE ISSCC*, Cut-and-paste organic FET customized ICs for application to artificial skin, 1:288–529, 2004.

[153] M. Watanabe et. al., *Proc. SPIE*, Development and evaluation of a portable amorphous silicon flat panel x-ray detector, 4320:103, 2001.

[154] Y. Vygranenko, P. Louro, M. Vieira, J. H. Chang, and A. Nathan, *J. Non Cryst. Solids*, Low leakage current a-Si:H/a- SiC:H n-i-p photodiode with Cr/a-SiNx front contact, 352:1837–1840, 2006.

[155] J. H. Chang, T. Tredwell, G. Heiler, Y. Vygranenko, D. Striakhilev, K. H. Kim, and A. Nathan, *MRS Symp. Proc*, Physically based compact model for segmented a-Si:H n-i-p photodiodes, 1066:A18–08, 2008.

[156] N. Safavian, Y. Vygranenko, J. H. Chang, K. Kim, J. Lai, D. Striakhilev, A. Nathan, G. Heiler, T. Tredwell, and M. Fernandes, *MRS Symp. Proc*, Modeling and characterization of the hydrogenated amorphous silicon metal insulator semiconductor photosensors for digital radiography, 989:A12–02, 2007.

[157] L. E. Antonuk, J. M. Boudry, Y. El-Mohri, W. Huang, J. H. Siewerdsen, and J. Yorkston, *Proc. SPIE*, Large-area flat-panel amorphous silicon imagers, 2432:216, 1995.

[158] R. A. Street, X. D. Wu, R. Weisfield, S. Ready, R. Apte, M. Nguyen, and P. Nylen, *MRS Symp. Proc*, Two dimensional amorphous silicon image sensor arrays, 377:757–766, 1995.

[159] R.L. Weisfield, *IEEE IEDM*, Amorphous silicon TFT X-ray image sensors, 21–24, 1998.

[160] R. B. Apte, R. A. Street, S. E. Ready, D. A. Jared, A. M. Moore, R. L. Weisfield, T. A. Rodericks, and T. A. Granberg, *Proc. SPIE*, Large area, low-noise amorphous silicon imaging system, 3301:2–8, 1998.

[161] M. Maolinbay, Y. El-Mohri, L. E. Antonuk, K.-W. Jee, S. Nassif, X. Rong, and Q. Zhao, *Med. Phys*, Additive noise properties of active matrix flat-panel imagers, 1841–1854, 2000.

[162] J. Lai, Y. Vygranenko, G. Heiler, N. Safavian, D. Striakhilev, A. Nathan, and T. Tredwell, *MRS Symp. Proc*, Noise performance of high fill factor pixel architectures for robust large-area image sensors using amorphous silicon technology, 989:0989-A14–05, 2007.

[163] J. L. Sanford and F. R. Libsch, *Proc. SID Symp. Dig. Tech. Papers*, TFT AMOLED pixel circuits and driving methods, 10–13, 2003.

[164] V. Vaidya, S. Soggs, J. Kim, A. Haldi, J. N. Haddock, B. Kippelen, and D. M. Wilson, *IEEE Trans. Circuits and Systems I: Regular Papers*, Comparison of pentacene and amorphous silicon AMOLED display driver circuits, 55:1177–1184, 2008.

[165] G. R. Chaji, P. Servati, and A. Nathan, *Electronic Letts*, A new driving scheme for the stable operation of the 2-TFT a-Si AMOLED pixel, 41:499–500, 2005.

[166] J. C. Goh, J. Jang, K. S. Cho, and C. K. Kim, *IEEE Electron Device Lett*, A new a-Si:H thin-film transistor pixel circuit for active-matrix organic light-emitting diodes, 24:583–585, 2003.

[167] S. W. Tam, Y. Matsueda, M. Kimura, H. Maeda, T. Shimoda, and P. Migliorato, *Proc. SPIE*, Poly-Si driving circuits for organic EL displays, 4295:125–133, 2001.

[168] S. H. Jung, W. J. Nam, and M. K. Han, *IEEE Electron Device Lett*, A new voltage-modulated AMOLED pixel design compensating for threshold voltage variation in poly-Si TFTs, 25:690–692, 2004.

[169] R. M. A. Dawson, et. al., *Proc. SID Symp. Dig. Tech. Papers*, A polysilicon active matrix organic light emitting diode display with integrated drivers, 438–441, 1999.

[170] Y. He, R. Hattori, and J. Kanicki, *IEEE Trans. Electron Device*, Improved a-Si:H TFT circuits for active matrix organic light emitting displays, 48:1322–1325, 2001.

[171] S. J. Ashtiani, P. Servati, D. Striakhilev, and A. Nathan, *IEEE Trans. Electron Device*, A 3-TFT current-programmed pixel circuit for active-matrix organic light-emitting diode displays, 52:1514–1518, 2005.

[172] J. Lee, W. Nam, S. Jung, and M. Han, *IEEE Electron Device Lett*, A new current-scaling pixel circuit for AMOLED, 25:280–282, 2003.

[173] Y. Lin and H.-P. D. Shieh, *IEEE Trans. Electron Device*, A novel current memory circuit for AMOLEDs, 51:1037–1040, 2004.

[174] M. Ohta, H. Tsutsu, H. Takahara, I. Kobayashi, T. Uemura, and Y. Takubo, *Proc. SID Int. Symp. Dig. Tech. Papers*, A novel current programmed pixel for active matrix OLED displays, 34:108–111, 2003.

[175] T. Sasaoka, M. Sekiya, A. Yumoto, J. Yamada, T. Hirano, Y. Iwase, T. Yamada, T. Ishibashi, T. Mori, M. Asano, S. Tamura, and T. Urabe, *Proc. SID Int. Symp. Dig. Tech. Papers*, A 13-inch AM-OLED display with top emitting structure and adaptive current mode programmed pixel circuit (TAC), 384–386, 2001.

[176] S. Ono and Y. Kobayashi, *IEICE Trans. Electronics*, An accelerative current-programming method for AMOLED E88-C:264–269, 2005.

[177] S. J. Ashtiani, and A. Nathan, *IEEE J. Display Technology*, A driving scheme for active-matrix organic light-emitting diode displays based on current feedback, 5:257–264, 2009.

[178] S. J. Ashtiani, and A. Nathan, *IEEE J. Display Technology*, A driving scheme for active-matrix organic light-emitting diode displays based on feedback, 2:258–264, 2006.

[179] G. R. Chaji, and A. Nathan, *IEEE Trans. Circuits and Systems II*, A current-mode comparator for digital calibration of amorphous silicon AMOLED displays, 55:614–618, 2008.

[180] S. J. Ashtiani, and A. Nathan, *Proc. IEEE Custom Integrated Circuits Conference*, A driving scheme for active-matrix organic light-emitting diode displays based on current feedback, 2:258–264, 2006.

[181] S. Sambandan, and A. Nathan, *IEEE Trans. Circuits and Systems II*, Stable organic LED displays using RMS estimation of threshold voltage dispersion, 53:941–945, 2006.

[182] I. Pappas, S. Siskos, and C. A. Dimitriadis, *IEEE Trans. Circuits and Systems II*, A fast and compact analog buffer design for active matrix liquid crystal displays using polysilicon thin-film transistors, 55:537–540, 2008.

[183] S. Sambandan, and R. A. Street, *IEEE Electron Device Lett*, Self-stabilization in amorphous silicon circuits, 30:45–47, 2009.

[184] B. Gilbert, *Elec. Letts*, Translinear circuits: a proposed classification, 11:14–16, 1975.

Index

For Product Safety Concerns and Information please contact our EU
representative GPSR@taylorandfrancis.com Taylor & Francis Verlag GmbH,
Kaufingerstraße 24, 80331 München, Germany

Printed and bound by CPI Group (UK) Ltd, Croydon, CR0 4YY

01/05/2025

01858482-0002